MW00745286

Strategic Buying for the Future

Strategic Buying for the Future

Opportunities for Innovation in Government Electronics System Acquisition

Barry M. Horowitz

The Libey Business Library

LIBEY PUBLISHING
INCORPORATED
A Regnery Gateway Subsidiary

Washington, D.C.

Published in the United States by
Libey Publishing Incorporated
a subsidiary of
Regnery Gateway
1130 17th Street, N.W.
Washington, D.C. 20036

Distributed to the trade by
National Book Network
4720-A Boston Way
Lanham, MD 20706

Library of Congress Cataloging-in-Publication Data:
Horowitz, Barry M.
Strategic buying for the future: opportunities for innovation in government electronics systems acquisition / Barry M. Horowitz
p. cm. — (The Libey business library)
Includes bibliographic references and index.
ISBN 1-882222-04-0 (alk. paper)
1. United States—Armed Forces—Procurement. 2. Electronics in military engineering. I. Title. II. Series.
UC263.H66 1993 92-40304
355.6'212—dc20 CIP

Printed on acid-free paper.

Manufactured in the United States of America.

10 9 8 7 6 5 4 3 2 1

For my wife Sheryl,
and to my children, Hillary and Charles

Acknowledgements

Many people in MITRE have contributed directly to the material in this book and have helped me organize my ideas and analyses. I am grateful and appreciative of their efforts.

Dr. Karen Pullen, Ramond Fales, Virginia Day, and Elsie Fitt of our Cost Analysis Center provided all the data related to cost throughout the book and wrote Appendices 1 and 2. James Spurrier and Frederick (Chuck) Marley of our Microelectronics Center supplied the charts and supporting information about the changes and trends taking place in the commercial microelectronics industry.

Judith Clapp contributed in a major way to the chapter on software architecture, with advice from Dr. Richard Sylvester and Gerald LaCroix. Robert Carroll adopted extracts from his reports on a maintenance study for the U.S. Navy for use as an illustrative example in the chapter on modernization.

Roy Jacobus and David Baldwin located examples, photographs, and charts drawn from a variety of different sources and helped edit and compose the text. Alan Shoemaker made the arrangements to bring all of these efforts together and provided guidance and help at every stage.

I would also like to thank the people in MITRE who have pursued these topics since this book was completed and who have helped refine the ideas. They include Charles (Skip) Saunders, Ronald Haggarty, and Robert McCown, among many others. In the end, it will be they who bring these ideas into practice.

Contents

List of Figures

List of Tables

Preface

I have presented much of the material in this book before, mostly in the form of briefings to individuals in government and industry. Since that time, I have received a very large number of requests for reprints of my papers and for additional discussion, which indicates to me that the topics are of great interest in the electronic systems acquisition community. I decided that it would be helpful to pull together all of this related material into a book, as a convenient reference for the community. I also thought it would be useful within MITRE.

Chapter 1
Introduction

Innovation in government systems procurement is possible — and it is fast becoming a necessity.

When the U.S. Department of Defense (DoD) wishes to develop a new system or modify an existing system, it unleashes a cumbersome process permeated with rules, constraints, procedures, and delays. This rigid mechanism (designed primarily to thwart excessive zeal and to promote competition) seriously reduces innovation in development organizations and industry. Many members involved with the mechanism are frustrated by its convoluted nature, and often give up trying to press against the rules rather than press for improvement.

The recent astonishing changes in the world situation have greatly intensified the stresses on DoD. The failure of Communism and the weakening of the C.I.S. (formerly Soviet) military have removed a major part of the threat against which the U.S. military forces have been directed. The danger of an all-out nuclear exchange is now thought to be remote, and many aspects of warfare that formerly dominated the design of military hardware and software (nuclear hardening, massive communications blackouts, physical attacks against communication satellites, etc.) are no longer believed to be important. Our new challenge is flexibility: We must be able to get our forces quickly to any part of the world, and we must be prepared to face multiple enemies, any or all of whom may be equipped with the sophisticated systems that can be bought on the open market by anyone with enough money. Political realignment is so unpredictable that we literally do not know who tomorrow's enemies might be.

To further complicate DoD's difficulties, the recent war in the Persian Gulf has demonstrated convincingly that high-technology weapons and the corresponding command, control, communications, and intelligence (C^3I) support structures are devastatingly effective, and that we need to maintain our proficiency in producing ever "smarter" systems. Much of our recent progress has been based on advances in electronics, computers, and software technology. The progress of these technologies, as driven for the most part by commercial industry, is incredibly rapid, and almost any kind of electronics product is obsolete in only a few years. Thus, to stay current, DoD must somehow renew its operational equipment every few years with the very latest in technology, while at the same time retraining its personnel in operating the equipment.

While DoD is being forced to keep up the pace in the quality of its equipment, the money available for military use has been cut drastically, and the current downward trend in funding promises to continue for many years. The size of our military forces is being reduced substantially, and it is necessary to find new ways to cut the operations and maintenance cost of running our systems. Vast reorganizations are under way in the Services in an attempt to recognize and accommodate the new order. The number of "new starts" — where entirely new systems are developed — will drop dramatically, and attention instead will be focused on maintaining and upgrading our existing systems.

All these stresses, and many more, will cause difficulties for DoD, the Services, and the community that supports the development of military systems. However, pressure of this sort may motivate improvements in the underlying processes that control DoD, in an effort to streamline and become more efficient. It may also provide the opportunity for innovation, which is the heart of the matter.

I believe that innovation is still possible, despite the many constraints imposed by the cumbersome procurement system and the downsizing of the military. This book provides a number of examples showing how, through relatively minor changes to its present practices, DoD might greatly improve the effectiveness of its development mechanism. Through ideas such as these, it should be possible to develop and modify systems with less risk of failure and

MITRE's high-frequency communications technology program is an example of innovation in military communications.

with better adherence to cost and schedule promises. Furthermore, most of the suggested changes can be applied to individual development programs, requiring only that the program be supported in putting them into effect.

This book was written partly because the DoD community should find these specific ideas useful; it was also written to help the DoD community with the notion that innovation is indeed still possible. I have presented the concepts described here to numerous leaders in the Air Force, the Army, the Navy, and industry, and have generally received considerable interest. In some cases, the ideas are actively being pursued; in others, they are under consideration. I am convinced that many more such ideas can be formulated if we take advantage of the peculiar opportunities afforded by the rapid changes in the military threat, in technology, and in commercially available products, and if we apply our efforts to the military's problems that arise from downsizing its budget.

Innovation in developing or modernizing military systems, however, requires substantial experience. I am the president and CEO of the MITRE Corporation, a private not-for-profit organization that was created to serve the government. Most of the MITRE Corporation supports the Office of the Secretary of Defense, all the Services, and many of the intelligence agencies in the development of C^3I systems. During my 22 years with MITRE, I have been an active participant in hundreds of different development programs, and I have observed a tremendous variety in the methods used to procure systems. From these experiences, I have drawn lessons about which techniques work and which are inherently weak. Of course, MITRE is as bound by the prevailing DoD practices as anyone else. But the unusually wide variety of system-development experiences that we have encountered over the years places us in an excellent position to comment on the overall process and to develop insight into methods that might work better.

As might be expected, many of my examples will refer to the development of C^3I systems, which includes command centers and decision aids, surveillance radars, various passive sensors, and com-

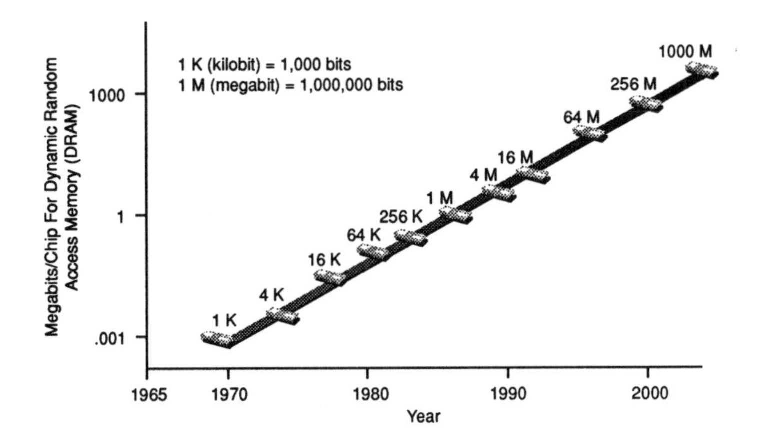

The pace of technology's growth is increasing exponentially; since 1970, the number of memory bits per chip has more than doubled every two years on the average. This trend is expected to continue well beyond the year 2000.

munication subsystems of all types (radios, satellite terminals, switching centers, etc.). While it is true that this class of system is not representative of all military procurements, it does involve a large number of challenges. Complex computer hardware and software are at the core of nearly every C^3I system; integration across numerous heterogeneous systems (the so-called "system of systems") is generally required, and the requirements for integration are often poorly formulated and subject to substantial evolution as the users learn new uses for the systems. Thus, while most of the examples are drawn from the development of C^3I systems, it is my intention that the ideas be applicable to nearly every aspect of military procurement and to much of non-military development as well.

The details of the specific military systems used as examples in the chapters to follow are not critical to the discussion, but it is helpful for the reader to know roughly what the systems are supposed to do. The Glossary at the end of this book explains the many acronyms and briefly describes the systems discussed. The credibility of my argument is enhanced if the reader happens to be familiar with one or more of the systems I use as examples, and the reader can obtain details on all the systems by consulting the standard library references to military products.

Chapters Two through Five deal with four essentially different topics, all linked by the common theme of innovation. Chapter Two observes that DoD must maintain the current effectiveness of many electronic systems (*e.g.,* warning systems; systems intended to project force in the event of conflict), but will have significantly less money for operations and maintenance of its equipment. The next two chapters focus on the seemingly intractable problems of software development and maintenance that are increasingly dominating the development schedule and life-cycle costs of most military systems. Chapter Three notes the difficulties inherent in our inability to deal effectively with the architectural issues in our software-intensive systems and offers a set of approaches to improve our skills and management of software architecture. In Chapter Four, we describe the complex issues surrounding the integration of commercial off-the-shelf (COTS) hardware and software with new custom-designed subsystems and with existing systems. Chapter Five was stimulated by my participation on a 1990 Defense Science Board study of modeling and simulation.

The Defense Satellite Communications System (DSCS III) is an important part of our nation's military system. Upgrades and maintenance will be essential to ensure its continued long-term future effectiveness. Courtesy of DoD.

The E-3A Airborne Warning and Control System (AWACS) has already proven its effectiveness for more than a decade. Continual system upgrade will allow it to remain viable well into the future. Courtesy of USAF.

The world of technology is changing at an ever-accelerating rate, and the world of politics seems to change almost as fast. Innovation in government systems procurement *is* possible — and more than that, it is fast becoming a necessity. We cannot ignore the challenges of rapid political realignments, since our potential enemies will have access to sophisticated C^3I systems of their own. Just as technology has become ever more streamlined and efficient, so too can the process of procuring that technology.

Chapter 2
Modernizing Electronics in DoD Systems

Industry-generated modernization focuses on government requirements while giving industry the challenge of proposing growth options.

Electronic systems are vital to Department of Defense systems and occupy a growing part of the budget. As a period of continuous and significant budget reductions begins, there is fresh debate about how best to manage the remaining funds. This chapter proposes an economic strategy for DoD modernization and logistics support of electronics, and it includes new approaches to funding, procurement, and development of electronics, as well as to field support. The proposal builds on selective activities and policies already under way within DoD, but broadens them so that they apply throughout the military.

Most aspects of the proposal pertain to DoD electronics in general. To make the discussion more specific, this chapter focuses on C^3 systems. A recent case study in which MITRE applied some of the ideas described here is presented at the end.

Impact of Reduced Budgets on C^3

This discussion assumes that the DoD budget will be significantly reduced in real terms over the next several years. Large budget reductions will have a profound effect on overall force size and readiness: Personnel levels will be reduced, few new major weapon systems will be developed, and money for the operation and maintenance of current weapon and C^3 systems will be cut. Present C^3 systems will be viewed as adequate, given the reduced threat of global war, and so there will be reduced interest in improving their performance.

For most of DoD, the cost of building and operating a system is roughly proportional to the size of the force that uses the system. In contrast, the cost of implementing and operating a C^3 system is closely coupled to the size of the *mission,* which in turn dictates the infrastructure of sensors, communication links, and command centers. This hardware and software infrastructure of C^3 systems can be called the *fixed C^3 plant,* and is analogous to the capital investment of a manufacturing industry.

This is particularly true of strategic C^3 systems: A major portion of the strategic C^3 infrastructure follows function rather than form.

After Desert Storm, the AWACS (background) and Joint STARS (foreground) aircraft returned to ESC's headquarters at Hanscom Field, Massachusetts to be displayed to the public. A Patriot missile launcher stands between the aircraft.

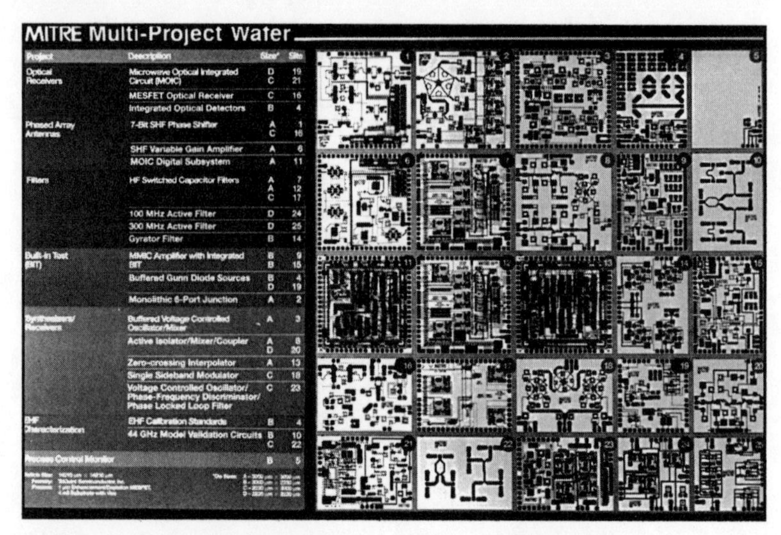

MITRE's Multi-Project Wafer is capable of performing 25 interrelated processes simultaneously. It is an example of trends in processing power and packaging, with ever-greater performance in ever-smaller space.

The number and types of warning sensors are primarily related to geographical coverage requirements; the number of major command posts depends on the degree of control rather than on the quantity of C.I.S. or U.S. warheads. The number and types of communication links required to support the strategic C^3 infrastructure are related to the geographic dispersion of weapons and sensors, as well as to various nuclear and electronic threats, such as electromagnetic pulse (EMP), disturbances of the ionosphere, and jamming. These factors increase the cost and complexity of the systems but are not closely correlated to size of either the U.S. or the C.I.S. strategic forces.

In fact, as the size of the deterrent force is reduced through treaties and budget cuts, C^3 connections with the weapons that are left must be even more dependable.

A reliable C^3 infrastructure is particularly important to new weapons systems, such as the B-2 bomber and the mobile land-based missile. For example, the B-2 needs C^3 support in attacking the enemy's strategic relocatable targets; to remain stealthy while penetrating enemy territory, it must not radiate radar or radio signals, and, therefore, is heavily dependent upon external C^3 systems to provide targeting information. Furthermore, the relatively small number of B-2 aircraft and their extremely high unit cost make the importance of tactical warning (a C^3 function) greater than ever, so that they can be quickly flushed from their bases under attack. Since budget cuts mean that the B-2 will operate from fewer bases, the number of B-2s at risk per base will increase. Similarly, mobile missiles need to be activated through a combination of strategic and tactical C^3 warning, and the post-attack role of these weapons in the deterrence equation dictates a correspondingly sophisticated C^3 system to control them.

Maintenance of C^3 systems is labor-intensive, requiring the attention of many thousands of skilled people. Although some of this effort is devoted to adjustments of equipment, most of it goes toward replacing defective or failed parts. The total annual cost of the parts (individual components, boards, or whole assemblies) together with the labor cost of replacing them is substantial.

If the number of maintenance workers and their level of training are reduced, then their ability to keep C^3 systems operating at peak performance will decrease — there will be more down-time, and the C^3 systems will be operating below specification more often. If the money allocated to spare parts is reduced, then failed parts cannot be replaced and systems will experience additional down-time. The cost of replacing parts will continue to rise gradually, exacerbating the spare-parts problem. This reduction in system availability will become an increasingly serious problem as weapons are removed from the active inventory (a result of arms limitations treaties). Higher efficiency will be required with fewer weapons, mandating even better performance from C^3 systems. The reduction of C^3 system dependability and availability runs counter to the often-stated objective of a higher quality, better controlled, and more flexible, albeit smaller force.

Crew members of the Multiple Launch Rocket System (MLRS) rely on effective command, control, and communications. Courtesy of DoD.

How Best to Respond to DoD Budget Cuts

It is clear that the availability of C^3 systems must be kept as high as possible to maximize the efficiency of smaller forces. Therefore, it can be argued that operations and maintenance budgets for C^3 systems should not be reduced proportionally with the reduction in force size.

It seems inevitable that cuts (perhaps deep ones) will be made to C^3 operations and maintenance in response to extreme overall pressure on the defense budget, and it follows that there will be a dangerous period in which the availability of our C^3 systems is reduced—where our strategic warning systems are not alert enough, or our tactical C^3 systems are not able to integrate our forces well enough. We need some way of gradually restoring high system availability in the face of long-term budget cuts, by changing either the C^3 systems or the maintenance and operations concept.

The number of people required to maintain a C^3 system depends largely on the number of failures that must be identified, located, and repaired each day, and on the time consumed by each repair action. A major part of the problem of degraded system availability arises when the size of the maintenance force is reduced, because some of the repairs cannot be completed in time, and combinations of critical failures are thus more likely to cripple the system despite its inherent redundancy.

If the reliability and maintainability of C^3 systems (expressed, for example, as the mean time between critical failures and the mean time to repair, respectively) could be improved, then the number of failures per day would decrease, and fewer maintenance workers would be required to keep the system operating at peak performance. An important approach to dealing with the problem of fewer maintenance workers, therefore, is to improve the reliability and maintainability of the C^3 systems.

As an approximation, assume that the availability of a C^3 system is directly proportional to the number of maintenance workers. Then reductions in personnel could be offset by a corresponding improvement in system reliability and maintainability. For example, a decrease of maintenance staffing by one-third would require an offsetting 50 percent increase in the success rate of repairs (2/3 x 3/2 = 1), which might be achieved by a 50 percent increase in reliability.

In practice, the relationship between system availability and the number of maintenance workers is not proportional, because of cross-training limitations, travel time, and other inefficiencies. It is likely that reliability and maintainability will need to be improved by factors of two or three times the reduction in workers. In addition, due to the limited cross-training of the remaining maintenance workers, a dramatic improvement in fault detection and isolation will be necessary to effect the repairs.

In a period of low defense budgets, it is tempting simply to maintain the current C^3 equipment—the fixed plant—by replacing

Systems maintenance being performed in a MITRE laboratory by replacing circuit boards.

broken parts but otherwise leaving the system intact. It is important to recognize that this view is seriously in error — it is necessary to regularly refurbish C^3 systems *even if performance improvement is not needed or wanted.* C^3 systems wear out and become obsolete, just as fighter aircraft and tanks become obsolete. In the case of a C^3 system, refurbishment keeps the system from total collapse. Given that C^3 systems require a substantial amount of refurbishment, this activity should be steered so that the system's reliability and maintainability are improved. Refurbishment funds, wisely spent, can keep system availability high and reduce long-term expenditures for operations and maintenance.

High system availability cannot be achieved solely through keeping the equipment running smoothly; system workers need constant training through operational exercises. The funding for operational exercises must be continued despite defense budget cuts, even though these funds compete with maintenance funds.

Figure 2-1 shows an example of the effect of test and exercise on the availability (in this case, measured as connectivity with force elements) of a classified communications network. It can be seen that the availability increased dramatically during the early period of test, evaluation, and field engineering, but then dropped precipitously after these monitoring and exercising efforts ceased. Availability improved again when the exercises were resumed.

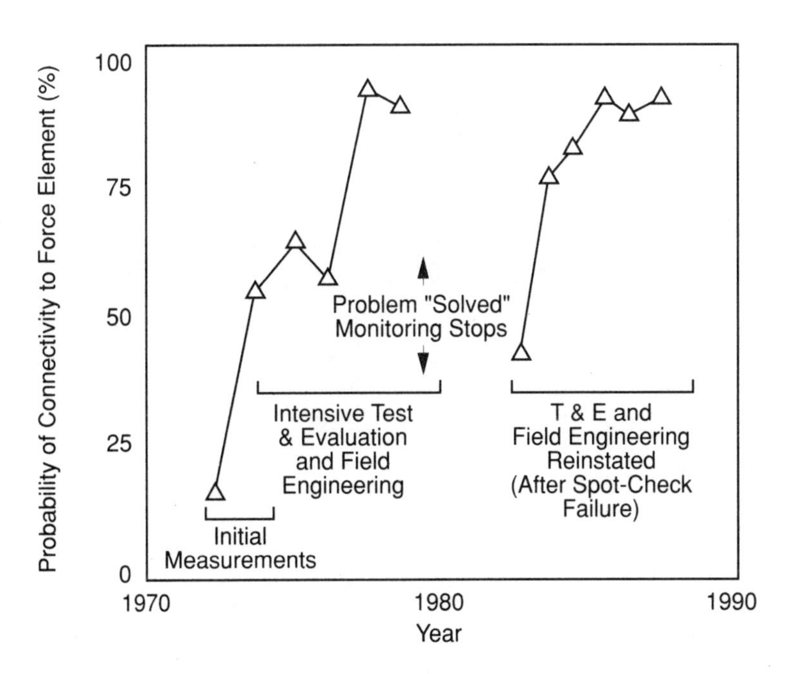

Figure 2-1. Historical test and exercise data for a classified communications network.

Four Approaches to Modernizing C³ Systems

The DoD will almost certainly reduce the number of maintenance workers and the funding for spare parts, resulting in decreased C³ system availability in the near term. Improvements in system reliability and maintainability can potentially offset these changes in workers over the long term. Obsolete parts no longer in production can be replaced in a program of continual system refurbishment and modernization; this activity can be the basis for incorporating the necessary changes in reliability and maintainability.

Given this overall view of the problem, there are four approaches that might be used to implement a suitable program of C³ system refurbishment:

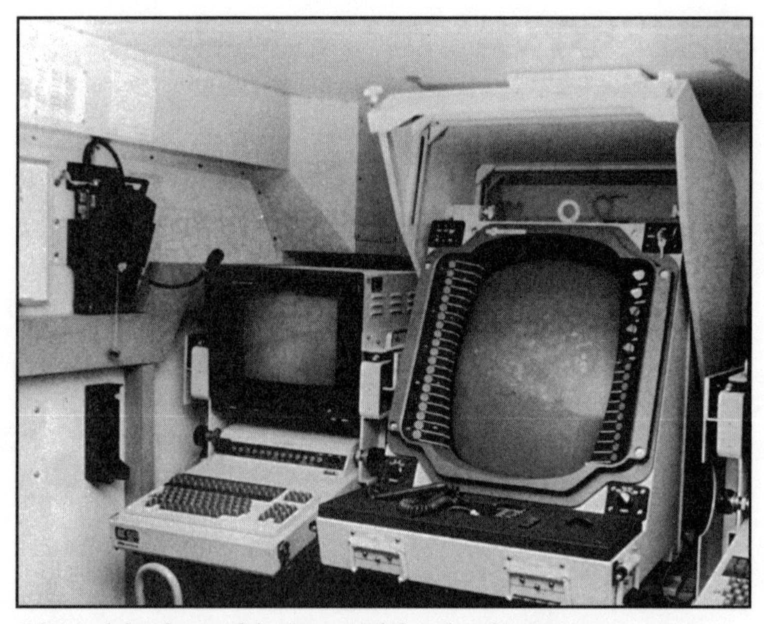

The modular design of the Army Mobile Subscriber Equipment System simplifies and reduces the cost of its modernization over its service life. Courtesy U.S. Army.

Approach	Method
1. Upgrade with No New Design or Development	Replace failed components with those of higher reliability.
2. Redesign of C^3 Systems for Improved Reliability and Maintainability (R&M)	Gradually refurbish C^3 systems as they become obsolete, while redesigning them to improve the R&M features. Set new reliability and maintainability goals, but do not improve performance in any way. Retain the performance specification from the original version of the system.
3. Development of New C^3 Systems for Improved R&M and Improved Performance	Initiate and execute development of new, higher-reliability, higher-performance C^3 systems through the current standard DoD acquisition process, beginning with the generation of a new performance specification.
4. Industry-Generated Modernization	Gradually refurbish C^3 systems as they become obsolete, and (as with Approach 2) redesign them to improve their reliability and maintainability features. Encourage industry to suggest cost-effective trade-offs for improved performance, but do not generate a new performance specification.

Approach 1 — Upgrade with No New Design or Development

This approach relies on the field-maintenance process to improve the reliability of system electronics by replacing failed parts with improved versions that have higher reliability. Field experience identifies components with relatively high failure rates, and the government can then procure replacement items with superior reliability, achieved by paying more for higher-quality parts and instituting better parts screening and stress testing.

The Maverick AGM-65B air-to-surface missile has been upgraded often since it entered service in 1972. Courtesy of DoD.

The straightforward field-maintenance process can provide some small improvements to overall system reliability (assuming the system has already met its original design goals for reliability), but cannot achieve large gains because it does not address the primary aspects of reliability. It cannot, of course, yield any improvement in fault isolation or any reduction of the time to repair.

As parts fail, maintenance workers call for the replacement of both obsolete and low-reliability parts. Replacing these parts requires design and parts-procurement efforts comparable to what the government does today for replacing obsolete parts. Given a significant reduction in maintenance staff, the demand for replacement parts must increase. The current replacement process for all DoD electronics has been estimated to cost more than $2 billion a year (see discussion under "Cost" later in this chapter). This estimate is probably too low, since the process is dispersed across acquisition, user, and support commands, making the comprehensive collection of data difficult.

System reliability improvements resulting from part-level upgrades are minimal. Electronics engineers recognize that, in most cases, the overall reliability of a system is dictated by its fundamental design rather than by the reliability of individual parts. Figure 2-2 focuses on the effects of improving a few parts whose reliability is especially poor; it shows the theoretical improvement in system reliability, as a function of the total number of parts in the system, when various combinations of bad parts are replaced. All of the system's parts are assumed to have the same reliability, except for certain bad parts. For curves A and B, the bad parts are assigned a reliability 10 times worse than all the others; for curves A' and B', the bad parts are 100 times worse than the others. For curves A and A', the original system has only one bad part, which is replaced by a part 10 times better than average; for curves B and B', the system has five bad parts, which again are replaced by parts 10 times better than

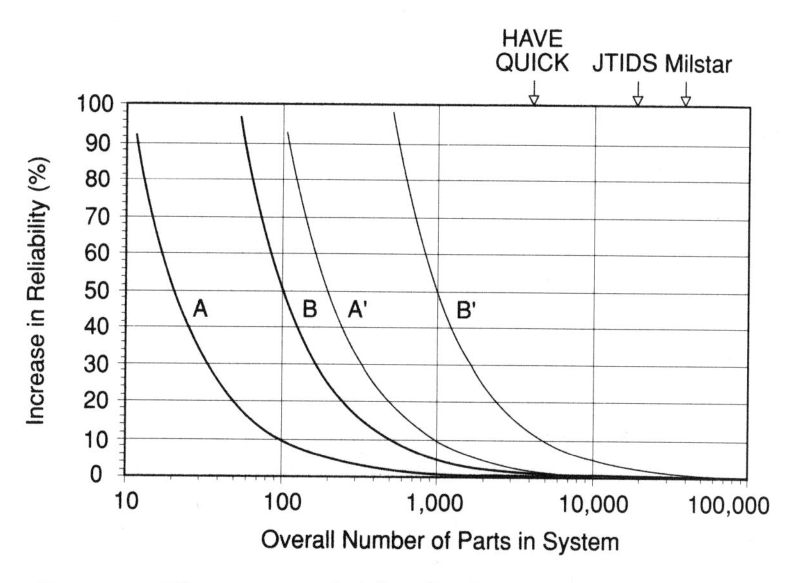

Figure 2-2. Effect on system reliability of replacing bad parts.

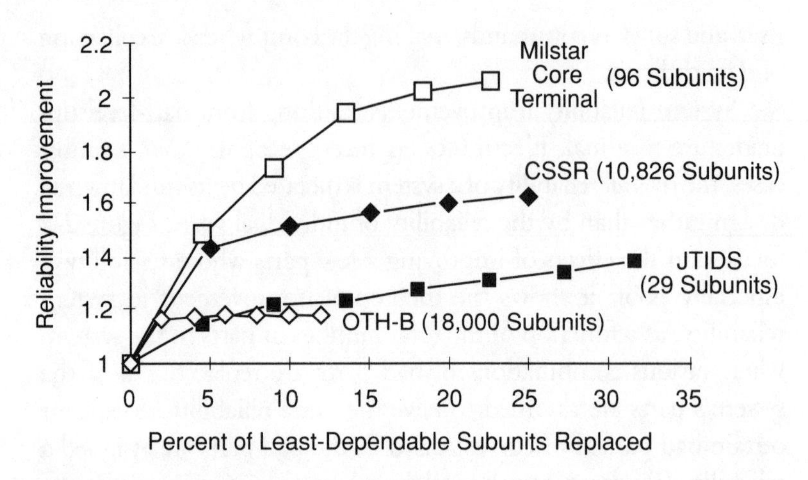

Figure 2-3. Improvement in system reliability by replacing subunits.

average. Three DoD communication systems, each with considerably more than 1,000 parts, are indicated.

Figure 2-2 indicates that improving the reliability of a few bad parts by factors of 100 or even 1,000 provides a very small increase in overall reliability. This theoretical result is strengthened by the observation that, in most mature systems, corrections have already been made for bad parts that might dominate overall system reliability.

Figure 2-3 shows what happens to the reliability of four C^3 systems when the reliability of the least-dependable subunits (as opposed to bad parts) is improved by a factor of three. System reliability does not increase until a significant amount of hardware is improved, and, in any case, does not approach a factor of three. The result simply illustrates that in practice the failure-rate performance of mature electronics is fairly evenly distributed across the assemblies.

The annual cost for parts replacement is increasing in general, because the obsolescence time of electronics is getting shorter and shorter. An electronic part becomes obsolete when there is no longer a demand for volume production. Once industry ceases mass production of the old part, there is no choice but to develop a new part (and often change the rest of the system as well), or to support the custom production of the old part at exorbitantly expensive unit rates. In almost all cases, it is better to redesign the part and take advantage of new technology with its improved reliability.

In the last 15 years, over 50,000 military parts have become obsolete. Figure 2-4 shows the dramatic increase in the number of discontinued or obsolete electronic parts (extracted from reports of the Government-Industry Data Exchange Program), with the number only partially through 1992 already surpassing the previous maximum by a substantial amount. Thus, one driver for increased parts-replacement cost is a growth in the number of parts no longer available from industry.

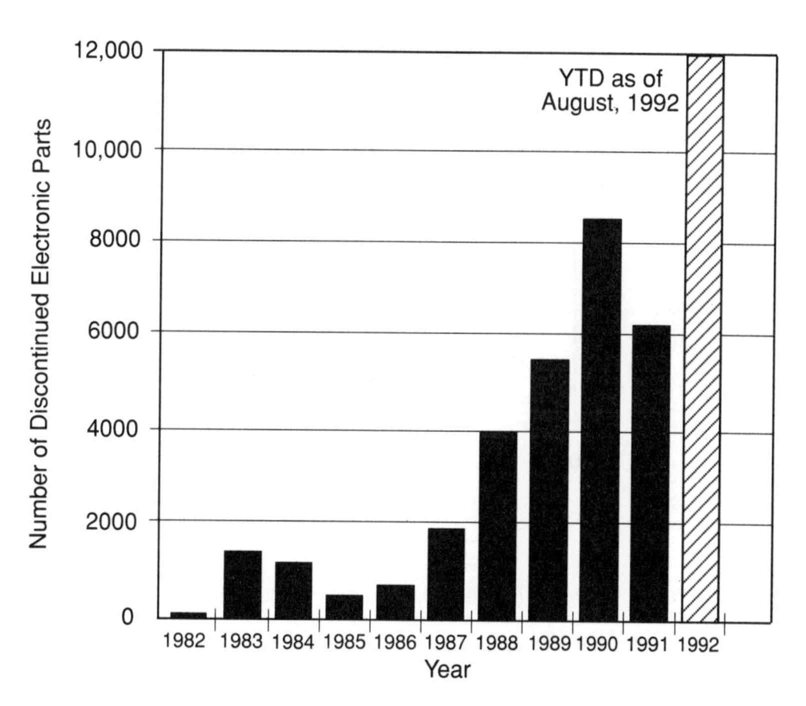

Figure 2-4. Electronic part discontinuation notices.

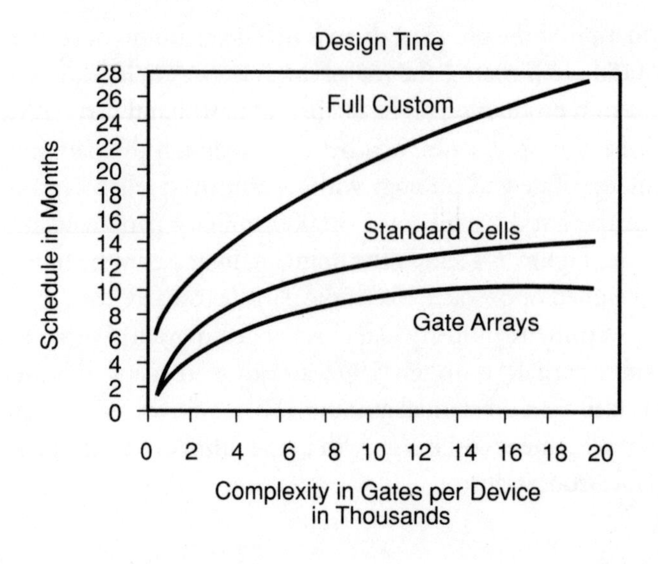

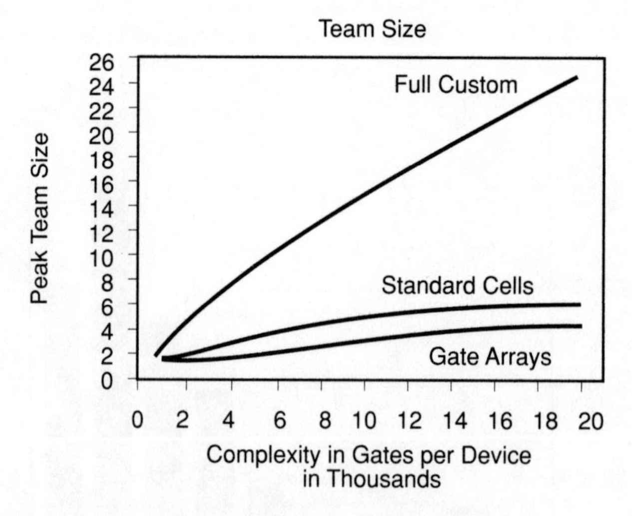

Source:
Studies in LSI Technology Economics III: Design Schedules for Application-Specific Integrated Circuits, D.E. Paraskevopoulos and C.F. Fey; IEEE Journal of Solid State Circuits Vol. SC-22 No. 2, April 1987, pp. 223-229.

Figure 2-5. Integrated circuit development effort versus circuit complexity.

The cost of redesigning parts also goes up with time because modern electronics have more integration per part, and therefore involve more complicated redesign. Because of the long development time for military systems, the effect of replacing more complicated parts is only now becoming apparent.

For example, the cost of redesigning integrated circuits grows as a function of the number of gates in the devices. Figure 2-5 illustrates how both the time required to design a chip and the peak number of workers on the design team increase with gate complexity. The curves for three different design approaches (gate array, standard cell, and full custom) show a maximum of only 20,000 gates; this value is probably consistent with levels of integration in many DoD systems (though today's levels of integration can exceed one million gates per chip).

Gate-array design is the least expensive, and full-custom design is the most expensive. For the case of a gate array consisting of 20,000 gates, the curves in figure 2-5 show about two staff-years of design effort, or about $250,000 per chip. If the DoD is forced to redesign thousands of chips every year, the total annual cost will be substantially higher than the traditional cost of replacement parts.

Design costs should drop somewhat over time as new computer-aided engineering tools are developed, but new design will still require new capital investment. System-related costs also increase as complexity increases. When the number of gates involved in a design is small, the chip designer can work independently and does not need to know much about the overall system design. As the number of gates increases, the number of system functions integrated onto the chip increases. Consequently, more chip designers are required and they must know more about the overall system design.

The result of following Approach 1 is predicted to be one of spending more and more money chasing in-field support problems at the parts and subunit level, as the C^3 system erodes in availability and dependability. This downward spiral is not a very rosy picture.

Approach 2 — Redesign of C^3 Systems for Improved R&M

Approach 2 is driven by parts obsolescence as well as by the need to improve reliability and maintainability; it calls for the modernization of whole systems, with the focus entirely on improvements to reliability and maintainability (and with no attention whatever being given to functional capability improvements). This proves to be much more rewarding, in a reliability sense, than the efforts of Approach 1 and is significantly less costly than Approach 3.

Modernization consists of the removal of some or all electronics from a fielded system, replacing them with a new set of electronics that furnishes exactly the same functional capability. When necessary, backplanes, connectors, wiring harnesses, power supplies, and even cabinets can also be replaced, so long as the external envelope of the

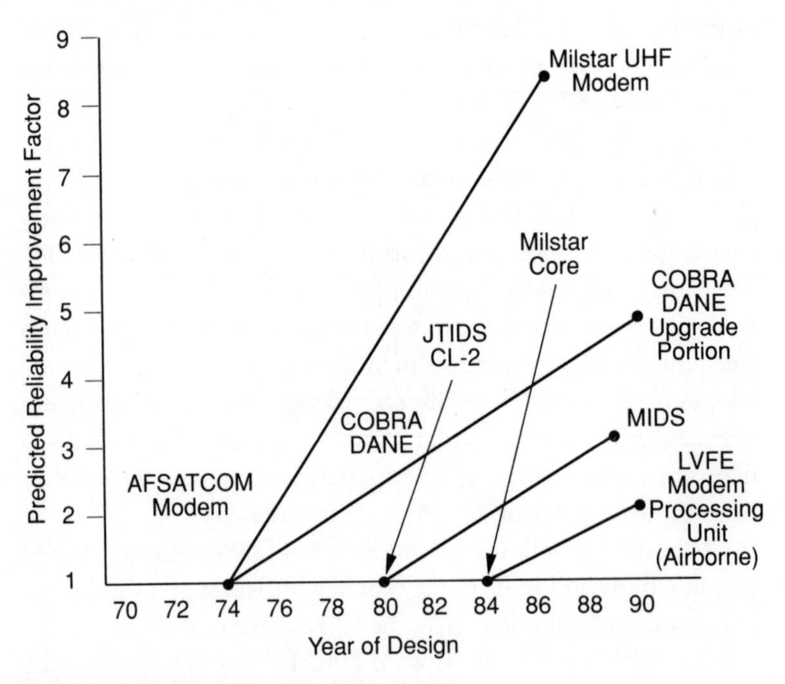

Figure 2-6. Predicted reliability improvement factors for redesigned systems.

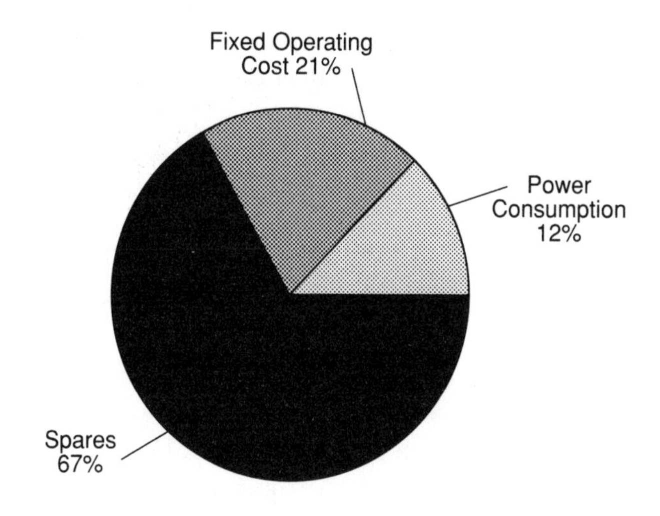

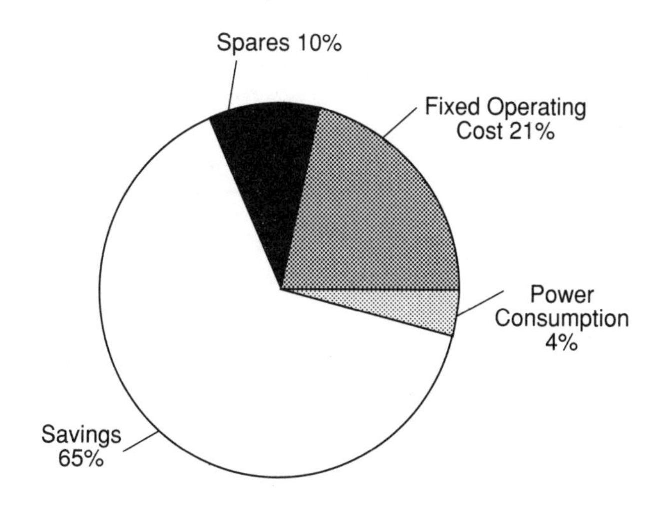

Source: Defense Communications Agency

Figure 2-7. Annual O&M cost for an SHF satellite terminal.

system is not altered and the electrical, mechanical, and human-factors interfaces remain the same. The original system-performance specification and the system test plan can be used in the development contract; the only changes are in the areas of reliability and maintainability.

Figure 2-6 shows the reliability improvements anticipated for four C^3 systems redesigned during the past few years. These systems were not redesigned specifically for reliability improvement, yet increases of two to eight times are expected due to the use of more modern technology. The slope of the curves suggests that average system reliability can be doubled every six years.

A recent example that demonstrates the benefits of equipment improvement is a Defense Communications Agency (DCA, now called Defense Information Systems Agency, DISA) activity aimed at modernizing a set of super high frequency (SHF) satellite communication terminals. "These terminals are designed around late

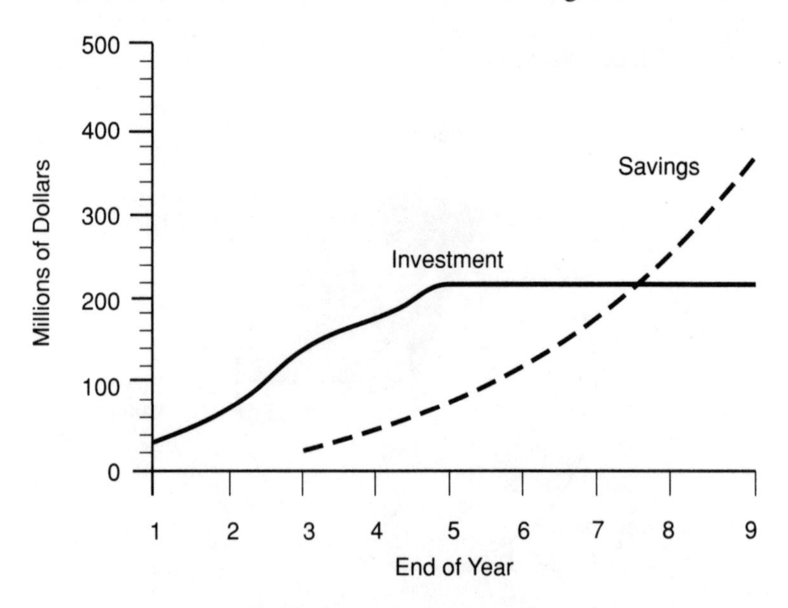

Source: Defense Communications Agency

Figure 2-8. Investment payback resulting from modernization of an SHF terminal.

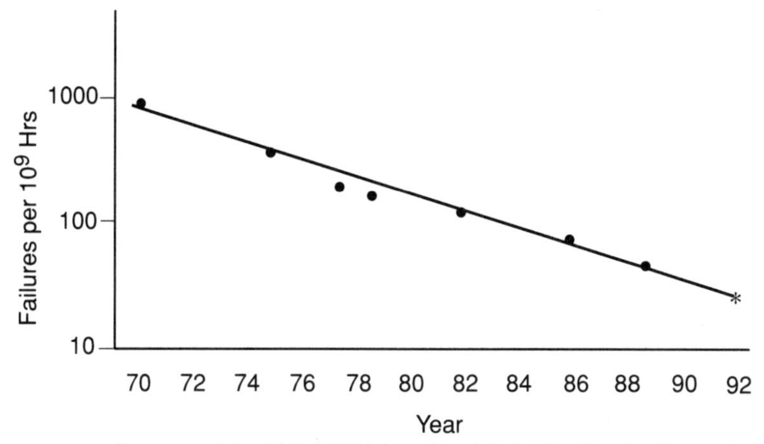

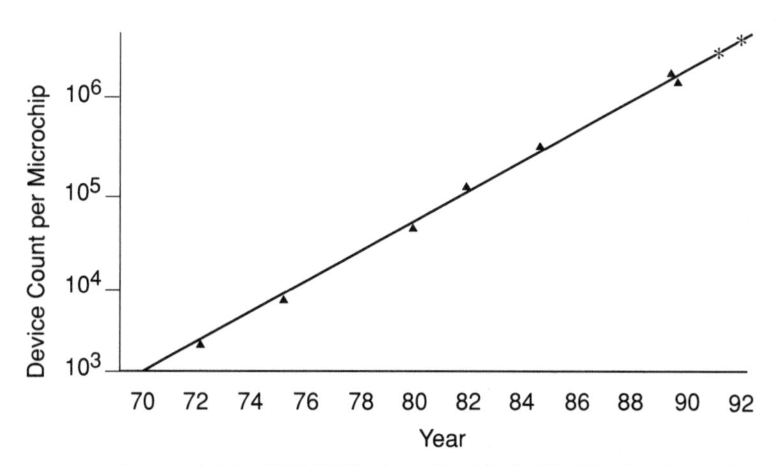

Figure 2-10. Integrated circuit device count trend.

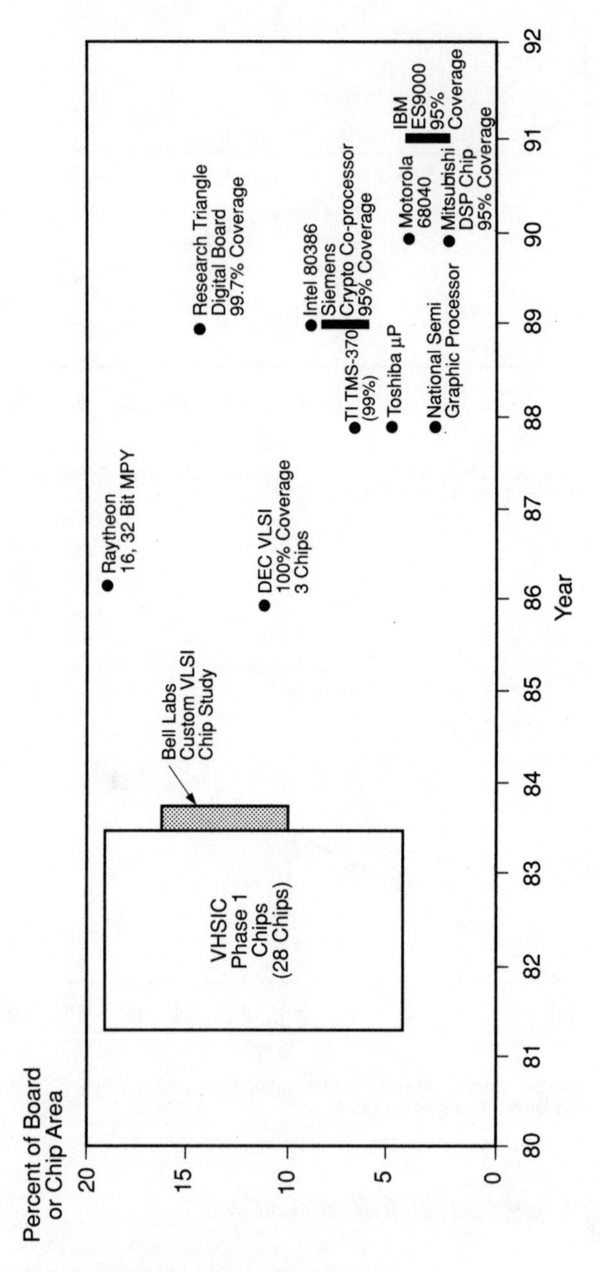

Figure 2-11. Percentage of board or chip area utilized for improved testability.

1960s technology. They involve the use of cryogenically cooled amplifiers and water-cooled transmitters, both of which are manpower intensive to maintain and require special skills. These terminals will begin reaching the end of their design life in 1991 and have already become extremely costly to operate and maintain. These operations and maintenance costs are projected to escalate dramatically without [a] modernization program . . ." (From a DCA briefing describing the benefits of modernization). Figure 2-7 shows the fraction of total annual cost for the terminals associated with operating cost, power consumption, and spares, referenced to 1992, for the two cases (a) where the terminal is not improved and (b) where the terminal has been upgraded through a modernization program. The savings are substantial.

Such modernization, of course, requires a development effort and a corresponding initial investment. Figure 2-8 plots the investment cost and the operations and maintenance savings, as a function of time, for the DCA example. The modernization effort is assumed to be a five-year program, conducted as a multiyear procurement, with equipment acquisition and program planning taking place during the first two years; installation (and corresponding cost savings from fielded systems) begins in the third year. After seven years, the savings exceed the accumulated investment of $206 million (a more rapid implementation could yield even greater returns). These improvements are due to improved manufacturing processes, greater integration, reduced power consumption, and increased use of built-in tests.

The potential for reliability and testability improvement through application of the latest electronic technology is illustrated in figures 2-9, 2-10, and 2-11. Figure 2-9 shows a decrease of nearly two orders of magnitude in single-chip failure rate over the last 20 years; figure 2-10 shows an increase of three to four orders of magnitude in the number of gates available in a package over the same time. These two factors, taken together, yield a potential increase of five to six orders of magnitude in the reliability per gate of digital integrated circuits over 20 years.

There is recent and intense interest in built-in (on-chip and on-board) test structures that enable individual units to assess their own health and status independently. Figure 2-11 shows the percentage of board or chip area devoted to test circuits for a variety of integrated-circuit products; all use less than 20 percent of the total useful area for test, and some products achieve very good fault-detection coverage.

Figure 2-12 illustrates results of a recent brassboard redesign and fabrication effort conducted at MITRE to modernize an integrated voice-coding and -decoding module in the Joint Tactical Information Distribution System (JTIDS). The principal goals were to reduce substantially the size and power requirements of the new electronics; the board area for the module was reduced by a factor of almost nine, saving much more area than would be needed for on-board fault isolation.

Increased fault-detection and fault-isolation capabilities can have a dramatic impact on system availability, allowing fewer maintenance support staff to find and fix failures more quickly. Since many different types of equipment are integrated into single platforms or

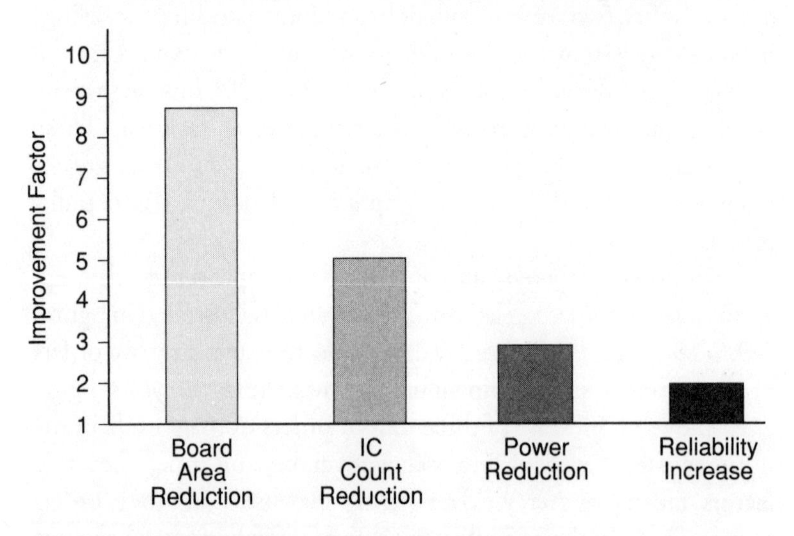

Figure 2-12. JTIDS Integrated Voice Module brassboard improvements.

at single locations, it is necessary to have high levels of automated diagnostic capability so that a few maintenance workers can provide adequate repair service for many equipment types.

In conventional DoD acquisition programs, it has been difficult to design high-quality built-in-test functions into systems because the hardware designs do not mature until late in a development project, and the software required to execute the function cannot be addressed properly until the hardware design is stable. As a result, efforts to develop built-in-test are usually late. Furthermore, the computer resources available for built-in-test are often inadequate because of unanticipated growth in other system areas; this forces the testing function to limit its coverage.

In modernization efforts, these negative pressures are greatly reduced, yielding a much improved built-in-test function. Furthermore, if a program of continued in-field upgrades to modernized systems were established, an improving built-in-test capability would be an important element for continued improvement.

What does it cost to modernize the electronics of a system, compared to the original development cost? Precise data are scarce, since almost all modernization programs actually involve significant improvements in system capability (which raises the cost). Some insight can be found from four C^3 programs that have recently been redesigned. The effective cost-reduction factors, defined as the ratio of the original hardware-design cost to the cost for redesign of the hardware, are shown in table 2-1. These approximate factors were derived by MITRE from a detailed examination of the cost histories of the original and the redesign development programs.

Table 2-1 Cost-Reduction Factors for Various C^3 Programs

Program Name	Cost-Reduction Factor
Milstar Low-Volume Terminal	2
Milstar UHF Modem	2
Have Quick Radio Upgrade	3
Microwave Landing System	3

Experience suggests that these cost-reduction factors would be even larger if the other program-specific requirements (such as volume reduction) were relaxed and the emphasis were placed solely on modernization.

Comparable cost reductions have been apparent in production costs and schedules. As an indication of this, some of the best electronics cost models show the weight of the end product to be closely correlated with procurement cost, almost independent of the type of product. Since modernization generally offers substantial

The Milstar project will enable the military services to communicate around the world. The system will serve all three Services in their development of ground-based, airborne, and shipborne terminals.

weight reductions, corresponding reductions in procurement cost can be expected.

Approach 2, the strategy of redesigning electronics for modernization only, is far superior to the do-nothing-new Approach 1, but introduces some of its own problems. First, initial costs are higher. Second, modernization offers almost irresistible temptations for performance improvement, especially since decision-makers are reluctant to allocate scarce funds without remedying at least some performance shortfalls in the original system. While this is not an unreasonable desire, it is one that can alter the entire process from problem *solving* to problem *creating* — that is, in attempting to improve the performance of the system as well as solving the problems of obsolescence and reliability, the process inevitably creates new problems. The impulse to improve performance may be so strong that Approach 2 may not, in fact, be practical.

This leads naturally to a third approach, in which the standard acquisition process is used to develop replacement systems that improve both availability *and* performance.

Approach 3 — Development of New C³ Systems for Improved R&M *and* Improved Performance

This approach represents the current development process. Improvements in performance are incorporated in the development effort through the standard requirements specification. Although modernization takes place, it is usually accompanied by the challenges of adding much more capability in a smaller volume, at a lower weight, and using less power. Development costs are high, schedules are long, and reliability can be lower than that of the less capable predecessors, certainly lower than when modernization alone is the goal. This undesirable set of events can occur even in the face of DoD policies that demand the achievement of double the reliability and half the time-to-repair for new systems.

In this approach, the usual process identifies user requirements, generates a specification (with reliability-related factors being prominent), and proceeds into development and production. As apparent from the cost-reduction factor of the four programs cited earlier, this kind of development is several times more expensive than modernization for increased reliability alone, and takes far longer. High cost and an extended schedule are due to the need to establish requirements, to create the specification, to identify and correct for the shortfalls of the specification during development and testing, to discover by trial and error the overly ambitious parts of the specification, and to accommodate the changes in requirements that occur during the long development schedule.

This process costs too much and takes too long, given today's financial pressures; furthermore, since there is great uncertainty in the future threat situation, the likelihood of establishing sound new requirements is even lower than usual. This exacerbates the problems inherent to the normal development situation, which are already formidable.

Approach 4 — Industry-Generated Modernization

As already noted, Approach 2 is attractive from a cost and reliability viewpoint, but may not be practical because of unwillingness to modernize for the sake of system availability alone. Approach 4 combines the attractiveness of Approach 2 with a strategy for adding new capability; at the same time, it controls development risk, cost, and schedule. Risk management is accomplished by motivating industry to provide performance improvements based on market forces rather than on government-specified requirements.

The normal process of development presents a problem: Users want to add new capability while modernizing their systems, but cannot be confident that budgets and schedules can be met. There are three basic methods of addressing this problem (other variants are both possible and desirable). The idea is to create a development process that is organized around the government's need to improve the reliability of C^3 systems and industry's ability to understand the relationships among possible new capabilities, time to develop, and production costs. One process option might work as follows:

Step 1 — The government decides to modernize particular C^3 systems, as in the case of Approach 2, with the objective of greatly improving reliability and maintainability. The original performance specification and acceptance test plan are used to describe the requirement.

Step 2 — The government augments this basic requirement with a list of performance-improvement wishes that industry is to treat as a market survey of customer interests.

Step 3 — The government, using a request for proposal (RFP), solicits industry to perform a system design and development effort that would include designing a basic set of internal electronics fundamentally directed at increasing reliability and maintainability, integrating into the design a set of compatible add-on performance upgrades judged by the bidder to be good trade-offs between cost, time, risk, and new capability, and building the basic modernized capability and testing to the test plan but not building the performance upgrades until directed to do so by the government.

Step 4 — The government selects two or more contractors to build the basic modernized system and then, based on demonstrations, possible upgrades, production costs, and so on, selects a producer. Alternatively, a single contractor can be selected with priced production options. In addition, during the lead-time period of production, the desired improvements can be developed for production consideration and installed later as field upgrades.

A different process option might proceed as follows:

Step 1 — The government provides industry with data on the support resources (spares, workers, technical orders, support equipment, and so on) for individual systems.

Step 2 — Members of industry are invited to propose two-phase modernization efforts of the type discussed above for any systems they wish, at any time they wish.

Step 3 — Consistent with a prescribed budget, the government funds the best-conceived efforts to the point of a production decision. Some systems may have more than one effort directed at their modernization.

Step 4 — The government makes the production decision based on price, quality, reliability, and the possible upgrades. During the lead time of production, the desired improvements can be developed for production consideration and installed later as field upgrades.

Yet another process option within the concept of Approach 4 is to award members of industry not only a contract for developing the modernized equipment, but also fixed-price, very long-term (10 to 20 years) life-cycle management contracts. These contracts can cover the efforts required for training military support workers in the field, repairing equipment, and developing the sequential upgrades needed to lower the cost of logistics support. In this arrangement, industry must trade off the schedule of modernization with the cost of retraining, the cost of obsolescence, and the dollar value of improvements in reliability, with the fixed-price contract arrangement providing the profit incentive. The Army has used this scheme on their mobile subscriber equipment program, and has documented their experience in a recent paper (Skurka, *et al.*).

The cost of the basic program is not much different from the reliability-only improvement program; however, this approach gives the government a chance to upgrade, and industry a chance to sell upgrades, using normal commercial market processes rather than the requirements-specification process. In addition, the designs can permit new consideration of the use of commercial parts, in place of military-standard parts, to reduce development time, lower costs, and increase the vendor base. This approach can also take advantage of the government-sponsored investments in Very High Speed Integrated Circuits (VHSIC) and Monolithic Microwave Integrated Circuits (MMIC), as determined by industry's judgment of applicability, risk, and competitive pressures.

Finally, this approach can move the DoD closer to a throw-away concept for failed electronics by achieving very high reliability, good fault isolation, and lower cost to replace than repair.

The proposed Industry-Generated Modernization process requires a change from the standard acquisition process to one that allows the government to select from a choice of products that are designed to meet different requirements. For the first process option described here, this type of modification can arguably be accomplished within the current methods by using the design improvements above the basic system as a general consideration in source selection and by raising the importance of general considerations in the source selection related to modernization. The second and third options require a more drastic departure from convention, but bring with them greater free-market orientation than the first option. From a first review of acquisition regulations, it does not appear that changes are required to permit this degree of departure from today's methods.

This concept has the same problem as the modernization approach — the need for large initial expenditures. This consideration is discussed later.

In summary, the first three approaches are either impractical or undesirable; the fourth is an attractive hybrid approach that may require some changes to current procurement methods.

Modernization of Software

Software is an important and complicating factor in DoD systems. The approaches for hardware modernization suggested earlier, particularly Approach 4, are valid for the software portion of modernization as well.

Older C^3 systems (such as our strategic warning systems), which are most in need of modernization, generally have less software than systems more recently developed, as illustrated in figure 2-13. Since

Software development for the Milstar Project.

the quantity of software in a system is often related to the difficulty and cost of modernization, high priority must be given to older systems that are not software intensive.

Communications and sensor systems tend to have less software than command-and-control systems. Figure 2-14 shows the ratio of software development cost to overall development cost in communications, radar, and command-and-control systems; the ratio is lowest in the communications system area and highest in the command-and-control and training areas.

Commercial software can often replace old custom software. This is true, for example, with display-generation software and data base management software.

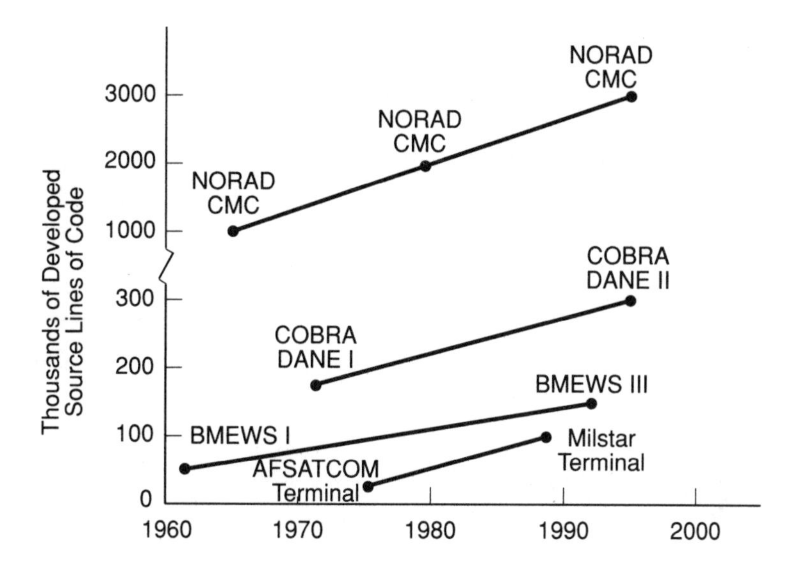

Figure 2-13. Growth of C^3 software size.

The capacity of new digital hardware is increasing at an astonishing rate, and it is now practical to implement much more capability than needed for the basic system. The large spare capacity readily permits optional software upgrades to be added after development of the basic system.

When appropriate, modernization can be carried out using Ada as a software language, thereby improving DoD's internal standardization position.

A broad-based modernization program can bring with it a focused effort at establishing common maintenance environments and tools to serve many different systems. This reduces training requirements and allows the same workers to maintain many different systems. Of course, some system-unique training will still be needed.

As the size of a software package increases, the difficulty of providing *exactly* the same functions on a different platform and in

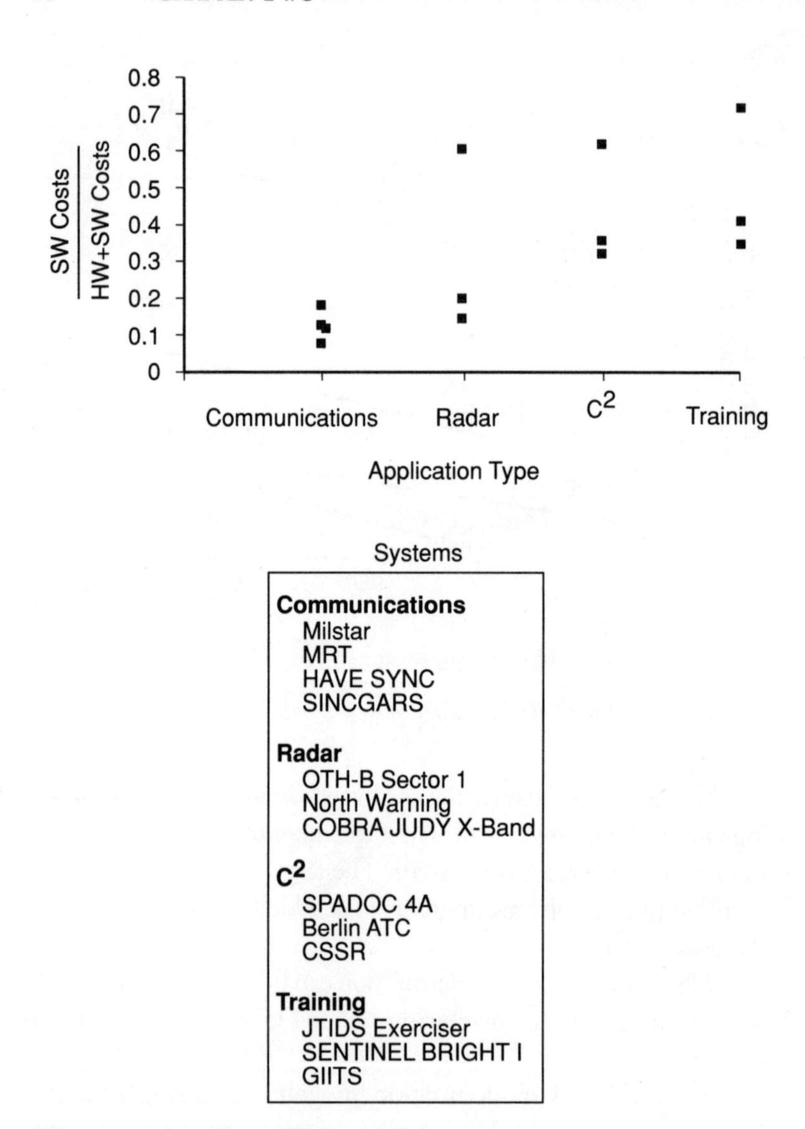

Figure 2-14. Relative cost of software development.

a different language increases dramatically. Proving that the replication of the functions of the old software is exact is often even harder.

It is not usually desirable to retain old, large software packages; they have typically grown in an unplanned manner, knowledge of the package's design rationale is sometimes lost (even though workers are still maintaining the package), and documentation is usually poor. Older software is often written in machine language, which is difficult to maintain and whose superior efficiency is no longer needed. A useful rule of thumb is that a software package should be totally redesigned if more than 20 to 30 percent of the original code needs to be rewritten.

Porting versus Emulation

The more modern systems in fixed C^3 plants are made up of highly integrated digital computer hardware, multiple processor configurations, and considerable amounts of software, with the software being the most crucial component. When the hardware becomes obsolete some time in the future, the cost-effective approach to maintaining the fixed plant must consider both hardware replacement and the impact of necessary changes to the software.

The two principal alternatives are "porting" and "emulation." Porting is the process by which old software developed on old computer hardware is re-hosted, or moved to a new computing platform. Since the new hardware may use a different instruction set from the old, and probably has a new software environment (*e.g.,* an operating system) as well, it is often necessary to make extensive changes to the old software routines for them to run at all. Emulation is an approach where the new hardware and its operating environment are made to appear exactly identical to the old system, so that the old software runs immediately without any changes whatsoever. The emulation of peripherals and older computers is common in commercial computing, and enables companies to upgrade hardware without having to modify all their software at once. Both old and new software is run on the new hardware; the system emulates its predecessor for the older software while running the newer, more recently modified software directly. Personal computers are often used to emulate earlier systems, permitting the continued use of older software while reducing maintenance costs considerably.

The difficulty of porting software varies in inverse fashion — command-and-control software is more easily ported than communications software, since the former is usually written in a high-level language and runs on standard commercial products. This is not always the case for communications systems (JTIDS required a special high-speed microprocessor, which in turn required a unique machine language for the software). Nevertheless, for high-volume software in communications and sensors, commercial processors are often used, thereby facilitating the porting of this software as well. A good example is the Air Force's Over-the-Horizon Radar program, where an array of standard Digital Equipment Corporation computers supports tracking and display functions. This tends to place a higher modernization priority on communications and radar systems, which on average are older than command and control systems.

At one extreme, a more modern, more supportable suite of hardware that is upward-compatible with the deployed hardware may be available. The old software can be ported to the new hardware

Surging air traffic places stringent demands on the national air traffic control system elements, such as this control center. Courtesy of FAA.

platform easily, with few if any changes to the software (unless specific timing relationships must be retained). To decide whether porting is the preferred solution, users must ascertain if the new hardware, with its projected support costs, will be less expensive over an appropriate amortization period than continued support of the old hardware.

The FAA recently upgraded the hardware in its air traffic control network while preserving its software investment. An analysis of the software determined that, except for one machine-language instruction, the software would run on a more modern IBM processor. All locations in the software using that one instruction were automatically identified and changed. The upgrade also required the addition and modification of about 10 percent of the 1.5 million lines of code. In addition, the FAA adopted a modern operating system; in the future, as the software is modified and maintained, it will work through the new operating system while the unchanged elements of software stay on the base machine.

At the other extreme (when there is no upward-compatible hardware and operating system), there are two combinable alternatives. The first is to modify or build a modern computer system to emulate the older devices; in this case, the mission software may have to be modified slightly to account for timing. The other alternative is to modify the software in order to move it to a new hardware platform. Once more the user must determine if, assuming the current capability is to be maintained, the expense of purchasing new hardware and porting the software is less expensive over an appropriate amortization period than the continued logistics support cost of the old hardware.

Striking the appropriate balance between hardware emulation and software porting must be done separately for each system, case by case, taking into account the risks and potential costs attendant on each approach. These factors favor porting of software to new computer hardware:

- Availability of an upwardly compatible family of computers (including compatibility of the new operating system)
- Cost-effective tools for translation from one instruction set to another

- A software package to be modernized that is simple, small, and independent
- Software written in a standard high-order programming language
- Standard interfaces
- Accurate documentation of the software
- Support staff members who are familiar with the software and its quirks
- No significant coupling of the software to peculiarities in the hardware
- No strong real-time constraints
- Availability of test cases to ensure correct and adequate performance and accuracy
- Availability of reverse-engineering tools for analyzing software-hardware dependence
- Having a small number of production units, which would make emulation of hardware architecture relatively expensive.

Other factors favor emulation of the old hardware by modern hardware:

- A software package that is large, complex, and integrated, and whose detailed implementation is no longer fully understood
- A large body of system users who are already trained in system use and whose retraining would be too expensive
- Particular timing requirements that are easy to meet, exceed, or precisely match using modern emulation techniques — by special microcode and firmware, through interpretive computer simulation, or a combination thereof.

It is clear that, when there is extensive software and where the system must continue to function during the transition to new hardware, emulation must be seriously considered as an option.

New computer tools are making this decision easier. For example, *reverse engineering tools* help identify those elements of the current software that depend on hardware and those that do not, simplifying the porting process. New *translation tools* can also assist in the porting process. *VHSIC Hardware Design Language (VHDL) representations* of hardware designs may make refabrication of complex systems more cost effective. Newer standards may provide us with compatible hardware (for example, Unix, Ada, and X-Windows). These new tools are not yet fully operational, but will eventually simplify many demanding evaluations considerably.

Improvements to Software during Modernization

Many of the systems that make up the fixed C^3 plant were delivered with some timing and sizing margins. To the extent that such margins are available and functions can be added without modifying the hardware and software architectural concepts, incremental performance improvements can be added to the software. Several factors can add substantial cost and risk to incremental improvement of the capability through additional software:

- Lack of good documentation and test cases for the software and system before improvement
- Lack of a knowledgeable software maintenance team either to advise or to make the change itself
- Multiple versions of the software at different sites (especially where site-peculiar adaptation is needed)
- A hardware/software design structure that is at odds with the desired incremental improvements (that is, the changes would seriously affect many of the software modules)
- No timing and sizing margins in the hardware (taking into account the need for an embedded testing resource).

Since software by its very nature is a flexible product, most software developers do not have enough discipline to avoid making at least small improvements — for example, by fixing software bugs. Such activity can lead to unwanted complications.

Most software support activity (over 80 percent) is devoted to incremental improvements to system capability. There are many lessons available through those activities on the extent to which modification to a current design is prudent; prominent among these lessons is the time it takes to become proficient with the software of a newly delivered system.

Cost

Every company has the problem of financing the replacement of its current plant to ensure continued competitiveness. Electronics-intensive plants have special problems due to the rapid obsolescence of electronic components and subsystems. Industry must make capability improvements in an evolutionary manner and simultaneously deal with obsolescence. In these situations, companies must set aside a steady flow of funds to modernize parts of the plant gradually, in concert with the rate of obsolescence. In computer-intensive plants, this might be a three- to five-year total replacement cycle; in other, less volatile situations, the replacement period might be extended to 10 years.

Table 2-2 shows some results of a 1991 survey of industry members involved in electronics. These companies own plants that depend heavily on support electronics (such as computer-aided design workstations). From the table, it is apparent that the ratio of large capital investments to current plant value varies between 7 and 26 percent. The obsolescence time is getting shorter and shorter, so the budgets for equipment replacement are likely to increase.

Table 2-2 Capital Investments by Selected Members of Industry ($ milllions)

Company	DEC	Intel	Nat'l Semi	CDC	IBM	GM	Honeywell
Capital Expenditures	$738	$948	$110	$33	$6,497	$7,300	$240
Total Value of Plant	$7,429	$3,644	$1,603	$384	$55,678	$68,432	$2,447
Ratio of Capital to Total Value	10%	26%	7%	9%	12%	11%	10%

Source: 10K forms and annual reports.

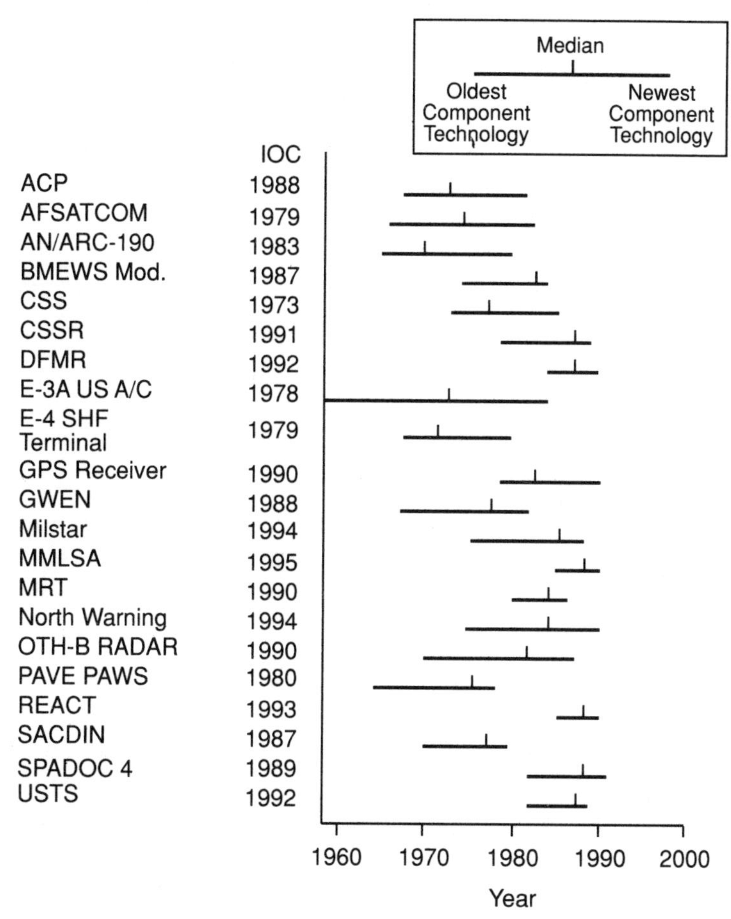

Figure 2-15. Age of components in C³ systems.

Over the next few years, a good portion of the DoD C³ plant will become obsolete. This is true even for the newer systems in the inventory. Figure 2-15 provides data on a selection of the C³ systems developed within the past twenty years. For each system, the figure shows the year of the system's Initial Operational Capability (IOC), the median age of the parts in the system, and the time span representing the oldest and newest parts in the system. Because of their long development time, the systems typically contain some very old parts—particularly when compared to the obsolescence rates for modern electronics.

Assuming that a ten-year replacement cycle is needed to avoid obsolescence in C^3 systems, then ten percent of the total replacement cost is a necessary annual budget for modernization. This basis for estimation is probably optimistic, because many electronic components become obsolete in much less than 10 years.

As an example, it is useful to determine the value of the strategic C^3 plant. One way to estimate this total investment is as follows: The accumulated strategic C^3 expenditures over the past 15 years are about $50 billion (dollar figures are in constant 1992 dollars; see appendix 2). Assume that about 75 percent of the C^3 plant is to be modernized over the next 10 years; further, assume that the cost for modernization is about half the original cost of the systems, based on the cost-reduction factors discussed under Approach 2. The result is about a $20 billion modernization effort. If this effort is spread uniformly over a 10-year period, then $2 billion per year is required for modernization. This is about one-third the current strategic C^3 development and acquisition budget (nearly $6 billion per year), and could grow to half the future reduced budget.

Thus, fixed-plant refurbishment represents a significant portion of the current development and acquisition budget for C^3. Nevertheless, the preceding discussion is a strong argument to support such an investment plan—in fact, when compared to the weapons development and acquisition budget, the fraction for C^3 modernization is small (between 5 and 10 percent), especially when the critical role C^3 plays for these weapons is considered. If such a funding strategy were applied to all DoD electronics, then roughly five to 10 times the calculated strategic C^3 modernization budget would be needed—about $10 to $20 billion per year.

These huge funding requirements are sobering, because it is clear that allocating this much money will be difficult in the coming years. The implication is that refurbishment of the fixed DoD plant will not proceed as it should, and that replacement will lag further and further behind the required pace, with a consequent loss in capability.

Decision-makers should recognize this trend and maintain their focus on allocating as much of the budget as possible to replacement and modernization.

Part of the difficulty is that straightforward replacement and simple improvements in system availability are neither glamorous nor readily perceived to increase war-fighting capability. This raises the question of whether there are additional benefits of modernization besides those considered thus far.

Additional Benefits of Electronics Modernization

While the discussion to this point has tended to focus on supportability and availability advantages, other benefits result from modernization. To make these benefits easier to visualize, we will use a tactical C^3 system as an example.

One of the most important attributes of a tactical system is its suitability for rapid deployment to remote parts of the world. Deployment can be measured by the amount of airlift capacity (cargo aircraft) needed to move the system and the number of workers required to operate and support the system. It is clear that modernization can offer significant improvements in the weight, power, size, reliability, and maintainability of a tactical system, but quantifying these gains can be difficult.

To make the potential gains more comprehensible, a brief analysis was performed for the equipment that would be deployed at a typical Air Force tactical communications node. This equipment, shown in figure 2-16, was developed in the late 1970s and early 1980s under the TRI-TAC program, and under the SHF satellite communication terminal program discussed earlier.

The communications node equipment is contained in seven tactical shelters, each of which is 12 feet long, 8 feet wide, and 8 feet high. For deployability, each shelter is mounted on a mobilizer and the shelter/mobilizer combination is pulled by a truck; 44 persons are required to operate and maintain the complement of equipment.

Equipment and people together weigh approximately 235,000 pounds and require six sorties by C-130 cargo planes to transport.

Considerable savings can be realized through modernization of the 1970s equipment in these shelters. A good example is the data-processing equipment in the technical control shelter. The present processor, memory, magnetic tape unit, and disk storage unit weigh over 200 pounds and occupy a rack approximately 19 inches wide and 72 inches high; the processor contains only 64,000 words of memory, and the disk storage unit has a capacity of only one million words. In addition, much of the equipment inside these shelters is interconnected manually through mechanical patch-panels.

Modernization can significantly reduce the bulk and weight of the communications equipment while improving its reliability and reducing the support resources required. The circuit-switch, technical-control, and multiplex shelters can be consolidated into one

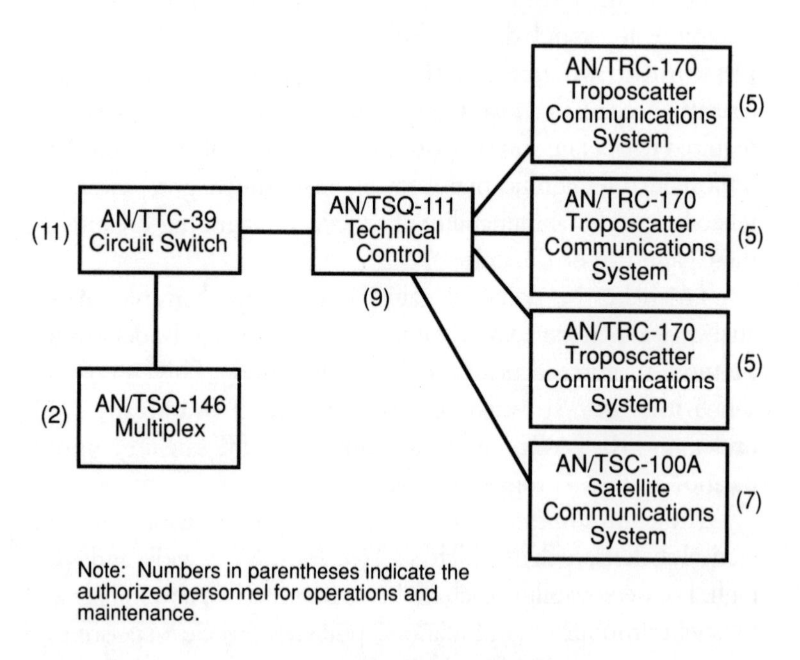

Figure 2-16. Typical Air Force tactical communications node equipment.

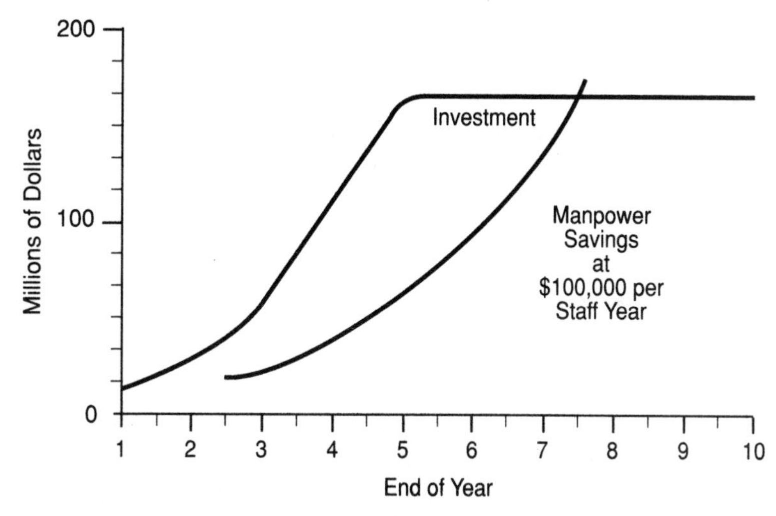

Figure 2-17. Investment cost analysis for modernized tactical communications node.

shelter; furthermore, this shelter can be mounted on its own vehicle, thereby eliminating the need for a separate mobilizer and truck. Operationally, it is desirable to leave each radio in its own shelter, but the equipment can be modernized and, as before, rehoused in a shelter mounted on its own truck. The resulting equipment is estimated to weigh only 64,000 pounds and require only three C-130 sorties for deployment. With each shelter on its own truck, there are the added benefits of increased mobility in loading and unloading the equipment and increased deployability once the equipment arrives in the operational theater. Finally, because there are fewer shelters and redundant equipment is removed, only 27 persons are required to operate the revised equipment configuration, a savings of 17 people (39 percent) per node.

Figure 2-17 shows a simplified investment cost analysis, illustrating how long it takes to recover the cost of modernizing and fielding the new equipment. It is assumed that 17 fewer operations and maintenance workers are needed for each node and that 20 nodes of equipment are modernized. The figure shows a break-even point at about seven years. Not included in this preliminary analysis are the savings from reduced airlift costs, lower fuel consumption of the modernized equipment, and the support costs of 40 shelters of equipment that were eliminated; factoring in these costs would increase the savings rate and make the break-even point occur sooner.

Getting Started

If it were decided to proceed with a steady effort of modernization following Approach 4, three DoD communities must be involved in the decision process. First, the operational community must decide which are the most mission-critical systems. Second, the logistics support community needs to single out the maintenance headaches. Third, the development community needs to evaluate the ease or difficulty entailed in modernizing each system (software plays a dominant role in this latter evaluation). The first two communities can represent the results of their deliberations by filling out the two-by-two logic matrix shown in table 2-3A.

Table 2-3A First Decision Matrix

	Many Maintenance Problems	Fewer Maintenance Problems
Higher Mission Criticality	Category 1	Category 2
Lower Mission Criticality	Category 3	Category 2

Systems that fall in Category 1 have the highest priority for modernization; systems in Category 3 might be deleted from the inventory. The development community can then evaluate the relative difficulty of modernizing systems in Categories 1 and 2 and estimate the effort and resources required. This input can be combined with the above matrix into a new two-by-two logic matrix (table 2-3B) to prioritize modernization actions further.

Table 2-3B Second Decision Matrix

	Easier to Modernize	Harder to Modernize
Category 1	Category 1a	Category 1b
Category 2	Category 2a	Category 2b

From this matrix, the first priority for modernization is established as Category 1a, which contains the systems that are the most useful, the biggest maintenance headaches, and the easiest to modernize. The remainder of the priority list can be determined based upon relative mission importance, cost, and modernization difficulty.

From the priority list thus established, an overall modernization program can be set up with an annual funding level determined as discussed earlier.

Once a particular program is selected for modernization, the development community can select a team to evaluate the basic and optional designs and to analyze the software and hardware trade-offs, as bases for conducting credible source selections. This eliminates many of the normal development-community functions, such as test planning, specification preparation, translation from operational requirements to technical requirements, and so on. However, the remaining team functions require detailed engineering knowledge of electronics design and software. Furthermore, the need to evaluate the optional design features credibly is crucial, since only the basic system is implemented at the time of selecting a producer.

A Case Study in Modernization

The concepts described above drew significant attention from many sectors within DoD, most notably within the Navy's Office of the Director of Space and Electronic Warfare. As a result, an effort was established to determine whether these concepts could be expanded to include a case study of the economics of modernization for a currently fielded Navy communication system — in short, to evaluate theory in the context of a real system.

The AN/SRC-16 High Frequency Communications Central and its derivative, the AN/SRC-23 transceiver, were identified as candidate systems demonstrating poor reliability and availability, and the Navy agreed that these systems were important to study. A year was spent, September 1990 to about August 1991, studying the problems and generating recommendations. After this work, the Navy decided to embark on a modernization program in September 1991, with MITRE acting as the systems engineer.

If detailed field-reliability data had been available on every part and subsystem of the Navy's inventory, and if the costs associated with maintaining the equipment had been identified in engineering terms, then this study would have been a straightforward exercise in optimization. As described below, the lack of these data greatly complicated the problem, and it was necessary to draw conclusions from partial data scattered among many different sources.

Identification of a Candidate System

The systems most in need of improvement must first be identified. Although modernization is expected to save money in the long run, the money required to modernize must be made available up front. The Navy has a limited budget for modernization, but it has thousands of systems that might benefit from modernization.

After the analysis of maintenance and cost data on more than 100 operational communications systems by using current Navy logistics data bases (specifically the Maintenance Material Management system, the Casualty Reporting System, and the Weapons System File available from the Ship's Parts Control Center in Mechanicsburg, Pennsylvania), nine indicators were developed and then used to rank the various communications systems. These indicators are listed in Table 2-4. Based on these rankings, the AN/SRC-16 High Frequency Communications Central was chosen as the subject of the study. Since major assemblies used in the derivative AN/SRC-23 transceiver (ranked tenth out of the 100 systems) are identical to assemblies in the SRC-16, we included this system in the study as well.

Table 2-4. Indicators for Modernization Candidates

Indicator Number	Indicator Description	Indicator Threshold
1	Total Corrective Maintenance Actions	> 1,000/year
2	Total Maintenance Man-Hours	> 15,000/year
3	Total Casualty* Reports	> 50/year
4	Total Repair Parts Cost	> $1.0M/year
5	Maintenance Deferrals	> 4/year
6	Parts Deferrals	> 4/year
7	Individual Maintenance Man-Hours	>200/ equipment/year
8	Individual Repair Parts Cost	> $2,500/ equipment/year
9	Individual Casualty* Reports	> 0.5/ equipment/year

• A "casualty" report is issued by the Navy when shipboard technicians cannot repair the system in a timely manner.

System Description

The SRC-16 and SRC-23 series of high-frequency (HF) radios have been aboard Navy ships for over 30 years. These radios are capable of sending digital data in a standard format called Link 11/ TADIL-A, used to maintain the tactical air, surface, and subsurface picture of a carrier battle group. The ability to maintain such a picture via HF over-the-horizon communications has been an important requirement for Navy battle-group commanders. This requirement has resulted in round-the-clock use of these systems whenever a ship is at sea.

The SRC-16, designed and built by Collins Radio Company (now a division of Rockwell International), was considered a state-of-the-art system when it was first deployed in 1959. The SRC-16 is designated a communications central because it includes communications patching, radio-frequency switching, and an array of below-deck multicouplers, in addition to four channels of receivers, exciters, and power amplifiers. The radio-frequency switching matrix allows other on-board HF gear to be patched through any of the system's 12 automatically tuned multicouplers. All together, the system occupies eight racks that are about 25 inches wide by 69 inches high by 27 inches deep. A block diagram of the AN/SRC-16 design is shown in Figure 2-18.

The SRC-23 is a single-channel transceiver that uses designs derived from, and very similar to, units in the SRC-16. The SRC-23 is housed in a single rack that measures about 14 inches wide by 72 inches high by 30 inches deep. Since two channels are required for Link-11/TADIL-A digital format (a primary and a backup channel), SRC-23s are always installed in pairs. Presently the Navy has 25 SRC-16 systems and 22 SRC-23 systems in operation. From Navy projections on expected ship decommissioning, we based our study on modernizing 22 SRC-16s and 21 SRC-23s.

The technology used in the SRC-16 and the SRC-23 is typical of the mid-1950s. Vacuum tubes are used in the signal path, and discrete solid-state devices are used in the bias and logic circuits. Band switching and tuning are accomplished with low-reliability electromechanical components. In the 30 years since the deployment of the SRC-16, significant advances in solid-state technology for communications equipment have been made. These advances have resulted in more reliable communications gear for the field; concurrently, the decreasing demand for vacuum tubes has made them increasingly more difficult to procure for spares. In some cases, the specific vacuum tubes originally required for the system are no longer manufactured, or are manufactured with significantly reduced performance. Similar commercial-grade tubes (versus more stringently tested military-grade tubes) or reconditioned tubes are often procured for spares, resulting in a shorter operational life. The advanced age of the systems and the unavailability of certain parts have resulted in rising maintenance costs and reduced operational

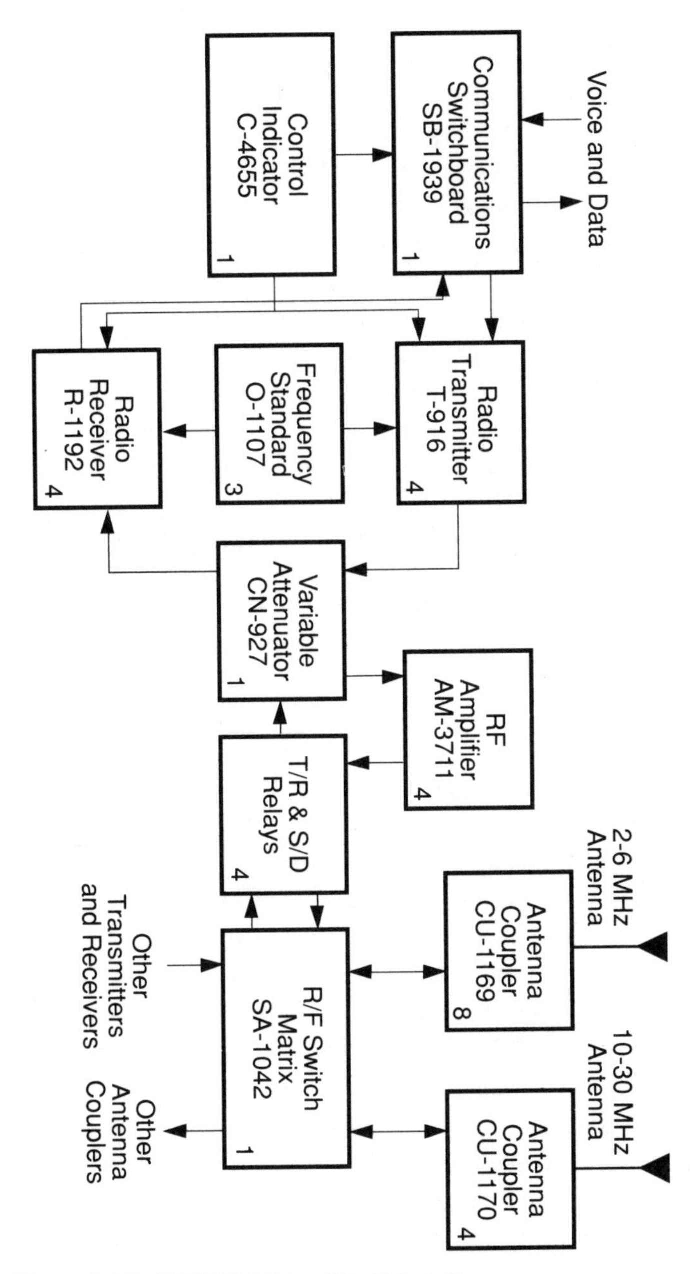

Figure 2-18. AN/SRC-16 simplified block diagram of the system design.

availability. The increasing expense of maintaining these systems made them ideal candidates for the modernization study, as the potential for savings was large.

The Navy had recognized the need for readiness improvements in these systems by nominating them to the list of systems monitored by the Detection, Action, and Readiness Tracking (DART) System, used by the Naval Sea System Command to focus management attention on those systems not meeting operational goals. The DART system recommended a number of initiatives to improve readiness, including improvements to technician training and positioning of spare parts. It was concluded, however, that only an engineering effort to modernize the systems would significantly improve readiness.

Reliability Analysis of the Targeted System

To formulate a modernization plan that maximizes reliability improvement while minimizing modernization cost, it was necessary to develop quantitative reliability models of the system. The reliability models allowed us to predict the impact of various modernization alternatives on overall system reliability, and ultimately on operations and support costs. Since failure rate was documented only at the system (vs. component) level, we had to devise alternative ways to use the available maintenance and logistics data.

Using the available data sources, we performed three analyses. First, we listed all the parts in order of demand and replacement rate. We manually sorted each of over 500 reported corrective maintenance actions performed on the SRC-16 and SRC-23 systems from July 1987 to July 1990 into one of four categories: the receiver/transmitter, power amplifier, multicoupler, and all others. (The receiver and transmitter were combined into one category because their assemblies were predominantly identical.) This gave us a rough basis on which to relate the effect of each component on the overall system reliability, as measured by the Maintenance Material Management data system. Third, we assumed that the replacement rate of a component was indicative of its failure rate. In this approach, the replacement rates of component parts (obtained from the Weapon

System File database) were combined to yield an overall model of the system failure rate.

From our analyses, we quickly determined that modernization of the receiver, transmitter, and power amplifier provided the most significant impact on overall reliability for both systems. Not coincidentally, these three units contain the bulk of the complex circuitry and nearly all the vacuum tubes in the systems. From our models we were also able to determine the point at which further reliability improvements in the three units offer diminishing returns at the system level. This provided a reliability goal for the modernization program and helped to minimize costs by ensuring that the modernized units were not over-specified.

To test our data analysis, we interviewed members of the SRC-16 and SRC-23 technical assistance team stationed at the Naval Electronic Systems Engineering Center in Portsmouth, Virginia. The technical assistance team gave us an overview of the major reliability problems in the SRC-16 and SRC-23 and described the shipboard technicians' typical troubleshooting procedures. These interviews validated our conclusions that the receiver and transmitter exhibited the major reliability problems, followed by the power amplifiers.

Both of our reliability estimation methods contain biases resulting from the shortcomings in the Navy's existing data-collection systems. The maintenance-action method yields a lower-bound estimate of the improvement in system reliability because many failures go unreported by shipboard technicians. The replacement-rate method leads to an upper-bound estimate for the improvement in system reliability, because the replacement rates were inflated by parts replaced in error (false failures). Using both methods together, we were able to bound our estimates of improvements in system reliability. These bounds were then used to provide upper and lower bounds for the expected cost savings. Table 2-5 shows the predicted improvements in the system failure rate for three scenarios: modernization of the receiver/transmitter only, modernization of the power amplifier only, and modernization of all three units. We estimated that modernization would increase the reliability of each unit by a factor of five. After reviewing the reliability of modern HF transceivers, we believe that this factor is easily achievable; furthermore, larger factors do not significantly increase the overall system reliability.

Table 2-5. Predicted System Failure-Rate Improvements

System	Modernization Option	Predicted System Failure Rate Improvements: Maintenance Action Method	Predicted System Failure Rate Improvements: Replacement Rate Method
SRC-16	Modernize Receiver/Transmitter	26%	49%
	Modernize Power Amplifier	9%	11%
	Modernize All Three Units	35%	60%
SRC-23	Modernize Receiver/Transmitter	30%	57%
	Modernize Power Amplifier	17%	14%
	Modernize All Three Units	47%	71%

Technical Approach

To estimate the cost of implementing a modernization, it was necessary to begin postulating specific technical approaches that might be taken. Given the variety of existing HF solid-state communications equipment in service today, the most economical and feasible engineering alternative was to use existing government off-the-shelf (GOTS) or commercial off-the-shelf (COTS) HF equipment if possible. Our technical analysis considered three principal options:

- Replace the entire rack that holds these units with a new one that houses modified off-the-shelf units

- Replace the units with new units that match in form, fit, and function
- Replace the internal circuitry of the existing units with new circuitry that matches in form, fit, and function.

The rack-replacement approach, attractive because of low development and procurement costs, resulted in prohibitively expensive installation costs. The third option, modernizing internal circuitry but retaining the existing chassis, proved unattractive because mechanical constraints imposed by the existing chassis layout would have prohibited the use of off-the-shelf modules.

We recommended a technical approach based upon the second option. Here, off-the-shelf signal-path modules are repackaged along with new interface modules into new equipment enclosures that match the existing receiver, transmitter, and power amplifier enclosures in form, fit, and function. The upgraded units will slide into the existing racks and interface with the other system equipment without requiring either mechanical or electrical shipboard modifications. This approach uses proven designs for low risk, minimizes nonrecurring engineering and qualification testing charges for low cost, and requires no modification to the existing equipment racks for easy installation. Thus, the modernization is a simple field change that can be performed by the shipboard technicians.

Cost Analysis

Having identified the modernization approach, it was necessary to document the total baseline annual cost of maintaining the systems in their current configuration and estimate the savings that could be realized based on the improved reliability and maintainability of the modernized systems. This proved to be no simple task because the accounting for each of the cost components is dispersed among the logistics, workers, and training communities; our success here was a unique achievement in itself.

Table 2-6 lists the annual operations and support costs for the current versions (unmodernized) of the SRC-16 and SRC-23 systems. Table 2-7 shows the total estimated annual operations and support costs for the modernized systems. These estimates were difficult because the desired data on individual parts failure-rates are not directly available, and it was necessary to infer this information from relatively gross maintenance data. A variety of inference methods were used in the study, yielding correspondingly different estimates of annual costs; for purposes of illustration, the estimates from only one method are shown here. Two options are presented in table 2-7: modernization of the receiver and transmitter only and modernization of the receiver, transmitter, and power amplifier.

In the shipboard labor category, there is no decrease for the modernized systems. This is because one full-time E-4 technician is currently programmed per ship, and we assumed that each modernized system would still require one full-time E-4 technician regardless of the extent of the improvement in system reliability. The decrease in recurring training costs is based upon an expected two-week reduction in the length of the technician's course. The cost of spares, consumable parts, depot labor, and technical assistance are decreased in proportion to the predicted reduction in failure rate.

Table 2-6. Average Annual Operation and Support Costs for Current Systems (thousands)

Category	SRC-16	SRC-23	Total
Shipboard Labor	$955	$912	$1,867
Training	292	292	584
Consumables	285	18	303
Spares	1,061	482	1,543
Depot Repair Labor	3,080	3,080	6,160
Technical Assistance	295	141	436
Total	$5,969	$2,538	$8,507

Table 2-7. Estimated Operation and Support Cost Savings for Modernized Systems (thousands)

Modernize Receiver and Transmitter	Operations and Support Cost Savings
Labor	$1,867
Training	500
Consumables	126
Spares	897
Depot Repair	1,894
Technical Assistance	214
Total Cost	$5,498
Annual Savings	$3,009

Modernize Receiver, Transmitter, and Power Amplifier	Operations and Support Cost Savings
Labor	$1,867
Training	500
Consumables	84
Spares	627
Depot Repair	1,442
Technical Assistance	160
Total Cost	$4,680
Annual Savings	$3,826

From table 2-7, we can clearly see that either of the two modernization options would provide significant savings in annual operations and support costs. These data alone do not tell us whether modernizing these systems is a good investment. The next steps were to estimate the cost to procure the modernization and then to perform a cost-benefit analysis. The equipment costs were estimated by pricing similar off-the-shelf equipment. The development costs were estimated through interaction with manufacturers of military HF communications equipment. In this way, we obtained realistic procurement cost estimates. Using our estimates for equipment cost and development cost, projected operations and support cost savings, and projected remaining equipment service life, we calculated total estimated savings, the payback period (point at which the savings outweigh the investment), and the net present value (10 percent discount rate) of the predicted cash flow stream.

Figure 2-19 shows the cumulative operations and support cost savings for modernizing the SRC-16 and SRC-23. The horizontal axis represents years from the modernization program start, and the vertical axis represents cumulative life-cycle cost savings resulting from the modernization. Development is assumed to occur in the first year of the program and deployment in the second year. Thus, each curve shows a negative savings in the first few years. By the third year all 22 SRC-16 and 21 SRC-23 systems are modernized. We also assume that the system has a remaining service life of 10 years. Figure 2-19 illustrates a few important points. Inclusion of the power amplifier yields a greater total savings over the life of the equipment,

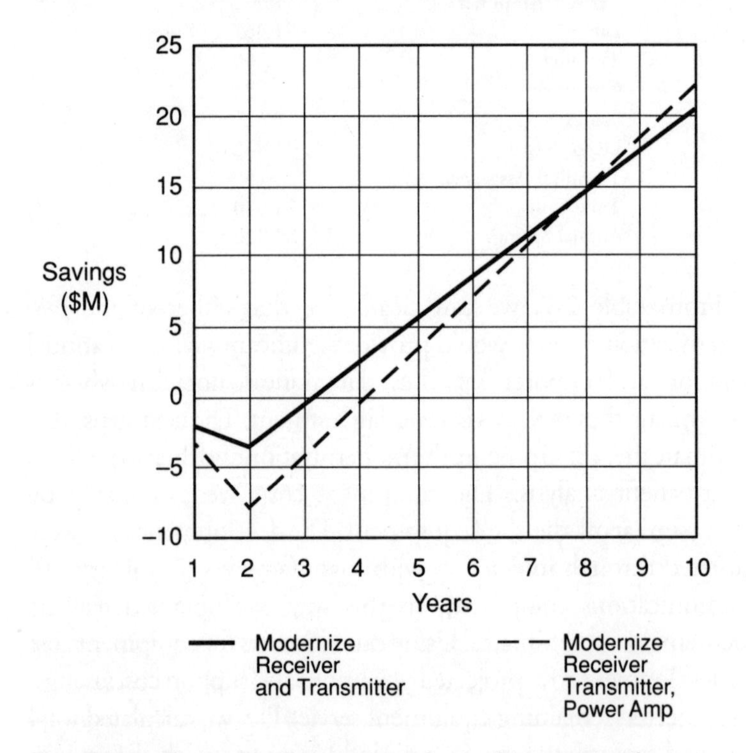

Figure 2-19. Projection of cost savings for modernization of SRC-16 and SRC-23.

but the payback period is longer (by one year) and the total savings does not exceed that of the receiver/transmitter-only modernization until at least seven years after the modernization program begins. This is because the unit cost of each power amplifier is almost three times that of a receiver or transmitter. The net present value analysis reinforced this point by predicting a lower return for the more comprehensive modernization plan.

Based upon our cost-benefit analysis, we recommended that the receiver and transmitter be modernized according to our recommended technical approach. A power amplifier modernization will enhance reliability and operational availability, but does not reduce the total system operations and support costs.

Results

We recommended the use of repackaged GOTS or COTS HF equipment to modernize the receivers and transmitters in the SRC-16 and SRC-23 communications systems. The failure rate of the modernized systems will be between 30 and 50 percent lower, resulting in a yearly operations and support cost savings of between $1.5 and $3.0 million for an investment of $4.6 million. The use of COTS equipment has several advantages, including proven reliability and reduced technical risk; GOTS has the additional benefit of an existing logistics infrastructure. With a number of existing, operational HF communications systems currently employing solid-state technology, there are a variety of contractors from which to choose. In addition, the modern equipment typically includes a number of features that could further reduce system maintenance requirements, such as improved testability. Based on the analysis and the forecasted payback period, the Navy is planning to embark on a project to replace the receivers and transmitters of those SRC-16s and SRC-23s on ships that were predicted to remain in service. The specification for the modernized receiver and transmitter does not include any new requirements and encourages the use of off-the-shelf modules.

White House, a prototype system for collaborative crisis management, was designed for maximum use of COTS components.

The AN/SRC-16 and AN/SRC-23 modernization program offered a unique opportunity to evaluate the concept of modernization on an operational system that plays a vital role in maintaining the Navy's C^3 picture. With declining defense budgets, the choice of modernizing existing systems in lieu of replacement is likely to become more and more common. The lessons learned from this study about the economics of modernization and the data requirements for reaching decisions should prove valuable to both the Navy in particular and DoD as a whole.

Summary

This chapter makes several points concerning the value and method of modernizing DoD electronic systems even in a time of lean budgets:

- Reduction in logistics support budgets decreases the dependability and availability of those C^3 systems whose functions are not correlated to corresponding reductions in force size. These include some very important surveillance, communications, and command-and-control systems that play an increasingly important role for new weapons systems such as the B-2.
- The cost to maintain an electronics system, as performed today, will increase significantly over the next several years as a result of the short lifetime of modern electronics, the greater design cost for the more complex integrated parts, and higher cost due to the need for more software support. Only small increases in system availability can be achieved by redesigning individual parts and subunits, so the decrease in availability brought about by cuts in support resources must be addressed at the system level.
- To solve this problem, an electronics modernization program should be initiated to systematically replace entire systems and subsystems as they age, with a 10-year period for total modernization of the inventory. This time constraint is driven by the obsolescence rates of electronic components. For the strategic C^3 inventory, an annual budget of about $2 billion is required (in 1990 dollars); this is about one-third of the current strategic C^3 development and acquisition budget. If this funding strategy is applied to all DoD electronics, roughly five to 10 times this annual budget will be required.
- The recommended method of development and acquisition, Industry-Generated Modernization, focuses the basic government requirement on increased reliability and reduced support requirements, but gives industry the chal-

lenge of proposing growth options in system performance. The government can buy these options later if funds permit.

- Such a program can ensure the dependability of the C^3 infrastructure that supports our forces, while permitting significant reductions in logistics support, staff power, and costs for electronic systems.
- The DoD should view its electronics equipment as a depreciating plant being written off at a given rate, with a corresponding investment rate to keep the plant viable. This is the normal cost of ownership that industrial firms must face, and it seems equally applicable to the U.S. government.

Recommendations

This chapter makes several recommendations about modernizing electronics in DoD systems during periods of lower budgets:

- The Office of the Secretary of Defense (OSD) and the armed services should develop plans for modernization along the lines of Approach 4, Industry-Generated Modernization. In addition, the services should present OSD with a set of those regulations, acquisition practices, and accounting schemes that need adjustment to permit a modernization effort to be conducted as efficiently as possible.
- Initial expenditures are required to get started. In the long term, a careful measurement process should be set up to measure the savings that result from using this approach. The measurements can then become the basis of a financial rationale for requesting and allocating out-year dollars for modernization.
- A research program should be established to focus on the problems of automatic diagnosis of systems, automated maintenance aids, and fault-tolerant system design. These efforts are needed to reduce the logistics support costs associated with highly reliable systems, since smaller maintenance staffs will be expected to handle less-frequent failures across a broader range of system types.

Chapter 3
The Importance of Architecture in DoD Software

We must work to introduce architectural improvements without disruption.

Our world is changing. The military threat to the United States posed by the Soviet Union for nearly 50 years is diminished, but there are new threats from rapidly evolving Third World countries that require rapid changes to military systems. Crises such as the recent events in the Persian Gulf highlight the need for flexible systems that can be changed quickly to meet the military's unanticipated challenges. In addition, the defense budget continues to be reduced — the government has less money to spend on systems.

The answer to this dual challenge — to make systems more flexible and to reduce the cost of defense systems — lies in the design of the digital system architecture, which includes the composition of hardware and software components, the structure that interconnects them, and the rules by which they interact. All too often, both government and industry focus narrowly on achieving the initial requirements for systems and give little thought to being able to adjust to what the system may be required to do five or ten years later, or to what happens as hardware may no longer be supportable or as advanced technology may become available for incorporation into the system.

Architecture design is the key to achieving the cost savings and operational flexibility inherent in digital systems. If the system is properly structured, then hardware components can be added or upgraded without expensive changes to the rest of the system. A good architecture allows a system designed to counter one threat to address a different threat through localized modifications to the software that change the functional capability of the system or allow it to interoperate with other systems. What is more, under the right circumstances, these changes can be made very quickly.

Chapter 4 discusses the critical role that system architecture plays in integrating commercial off-the-shelf (COTS) products into large

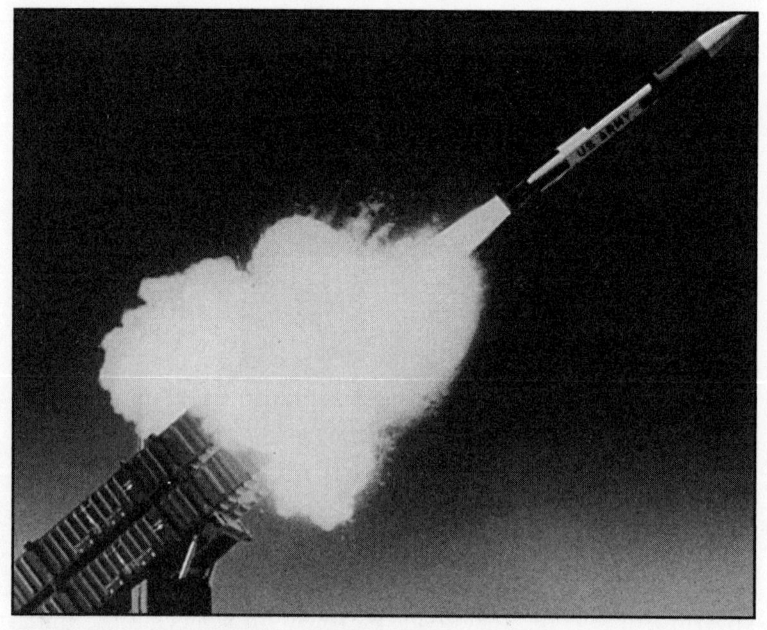

Patriot surface-to-air missile launched from a truck mounted launcher. Courtesy of DoD.

DoD systems, and expands further on some of the impediments created by our current development practices. In this chapter we take the view that for largely custom-designed systems, industry could be given the responsibility for selecting the system architecture. Chapter 4 argues that government is perhaps better suited to choosing the system architecture in situations where COTS software products will be dominant.

DoD Software: More Important — and More Expensive

The Value of Software

Software provides modern defense systems with a flexibility that cannot be achieved in any other reasonable way. Operation Desert Storm provides several excellent examples.

Patriot is an Army corps-level missile system primarily designed to counter aircraft. Given the inherent capability of the missile itself, the designers gave some thought to employing it to shoot down incoming enemy missiles. The Scud attacks during the war, however, focused everyone's attention on this threat with much more urgency.

Patriot's designers developed a new software package that increased the Patriot's effectiveness to counter the Scud threat. When radar tracks began to show that the Scuds were breaking up on reentry, the designers further tuned the package to recognize and attack the Scud warhead, and not the debris that accompanied it.

Without the modified software, Patriot would have been less effective. Yet the designers were able to implement this capability quickly and at a surprisingly low cost. No new missiles or radars were required. The software improvements could go to the war region in a briefcase.

Another example of this flexibility also involves Patriot, though at the much higher level of command, control, communications, and intelligence. To improve Patriot's ability to react to the Scud attacks, which proceeded at much higher speeds than the targets normally confronting Patriot, U.S. space satellites were redirected to watch for

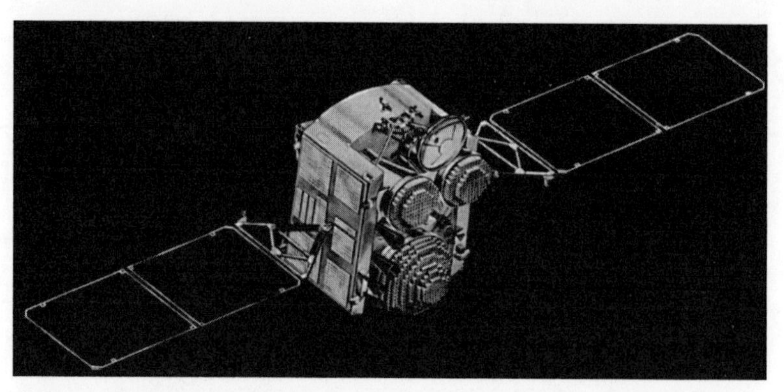

Software flexibility enables satellite systems such as ASTRO-DSCS III to handle unexpected communications challenges. Courtesy USAF.

Scud launches. When a launch was detected, the satellite relayed the targeting cues over a satellite link to the appropriate Patriot battery, leading to a successful interception. Minor software modifications permitted a network to be set up.

There are other, less dramatic, examples of the value of software's inherent flexibility that came out of Desert Storm. Navy attack aircraft had been set up for years to attack Soviet targets, either at sea or on land. A cassette data-tape provided the attack computers with the information they needed to launch their stunningly accurate attacks on targets that had only recently been identified.

Software was also the key to the effectiveness of Air Force jamming aircraft. Programmed for operations against Soviet-bloc radars, the jammers were faced with a mixture of Soviet, French, British, and Italian equipment. Software changes enabled the equipment to perform its task against this new threat far more quickly — and less expensively — than could have been done otherwise.

Precisely because of software's flexibility, the DoD is buying more of it for its systems and implementing functions in software that had previously been performed in hardware. Figure 2-13 shows this trend in a number of systems. For example, the latest version of the Cobra Dane radar system uses more software than did the original release, and the new Milstar terminal uses more software to perform more functions than did its predecessor, AFSATCOM. Desert Storm demonstrated that the flexibility software offers us is real and of great value to the military. It will become more so if we continue to have crisis scenarios that are harder to predict and cause us to apply our systems in unplanned ways.

The Cost of Software

Since the DoD has been buying more and more software, the total expenditure on software has been increasing, and software is expensive. With shrinking military budgets, we have to find ways to use more software and yet reduce its cost.

On average, two-thirds of what is spent on software is believed to be spent after the system becomes operational, during the maintenance phase, as illustrated in figure 3-1.

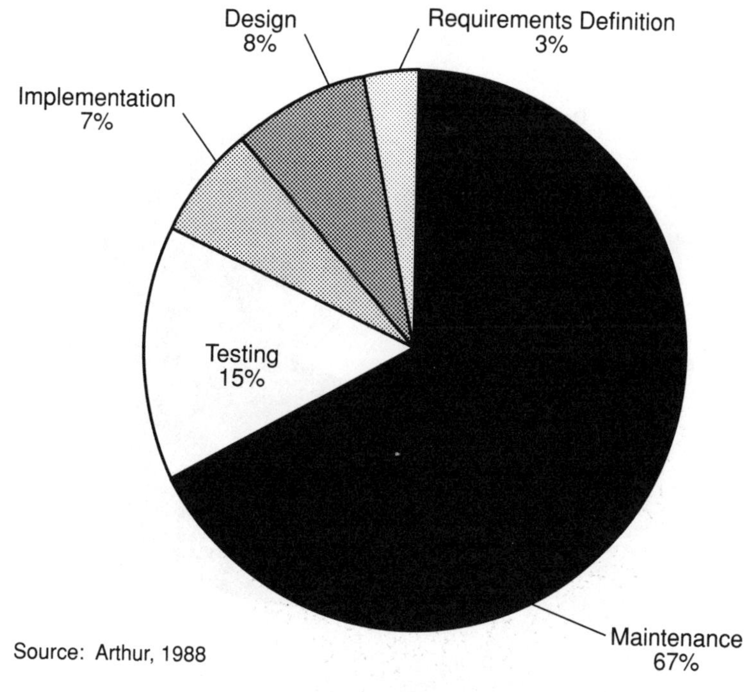

Figure 3-1. Software life cycle cost distribution.

Looking at the distribution of software maintenance activities is itself illuminating. About two-thirds of the software maintenance effort for a system is typically spent on modifying the original system to provide new capabilities and to add new technology — at least three times the effort spent on making repairs. Figure 3-2 confirms these ratios for an Army command and control system.

Taking these two sets of data together suggests that about 45 percent of the effort spent on software is used to change the system after it has been delivered. Experience also shows that we often spend part of the system development effort making changes in response to changing requirements (or requirements that become better understood). We probably spend more than 50 percent of our software effort to change the capabilities of a system over its developmental and operational lifetime.

If we can design software systems to take only half as much effort to modify, we can reduce the life cycle cost of the entire software

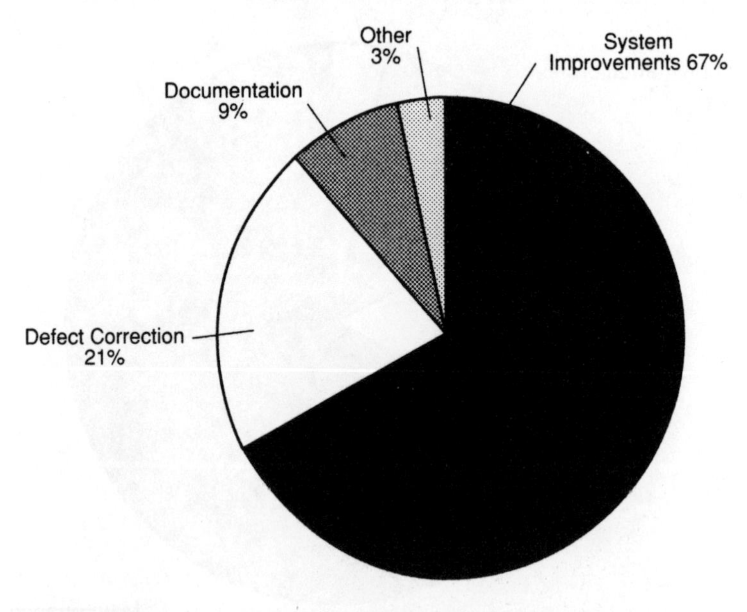

Figure 3-2. Software maintenance activities.

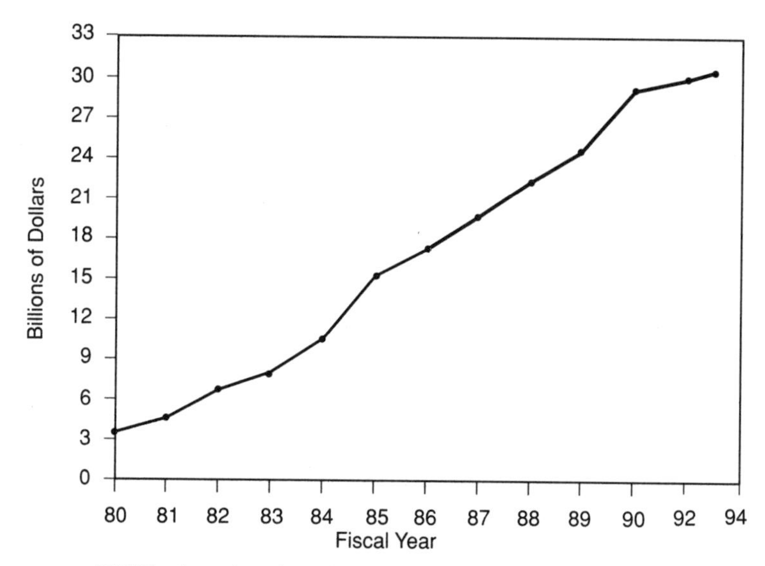

Figure 3-3. DoD software expenditures.

system by 25 percent. When applied to the total amount the DoD spends on software, this improvement can yield enormous cost savings.

While it is difficult to determine accurately how much the DoD spends on software, MITRE staff made a rough analysis that indicates the total amount to be approximately $30 billion per year (see figure 3-3). If we can in fact reduce the life-cycle cost of software by 25 percent, the total savings will range between five and eight billion dollars every year.

The example in figure 3-4 illustrates how these savings might be possible. Three thousand lines of new code were required to be added to a system of 50,000 lines. When the changes were made, the cost, time, and number of defects found in the delivered system were measured. As a separate exercise, the structure of the original software was improved and the desired changes were inserted into this new

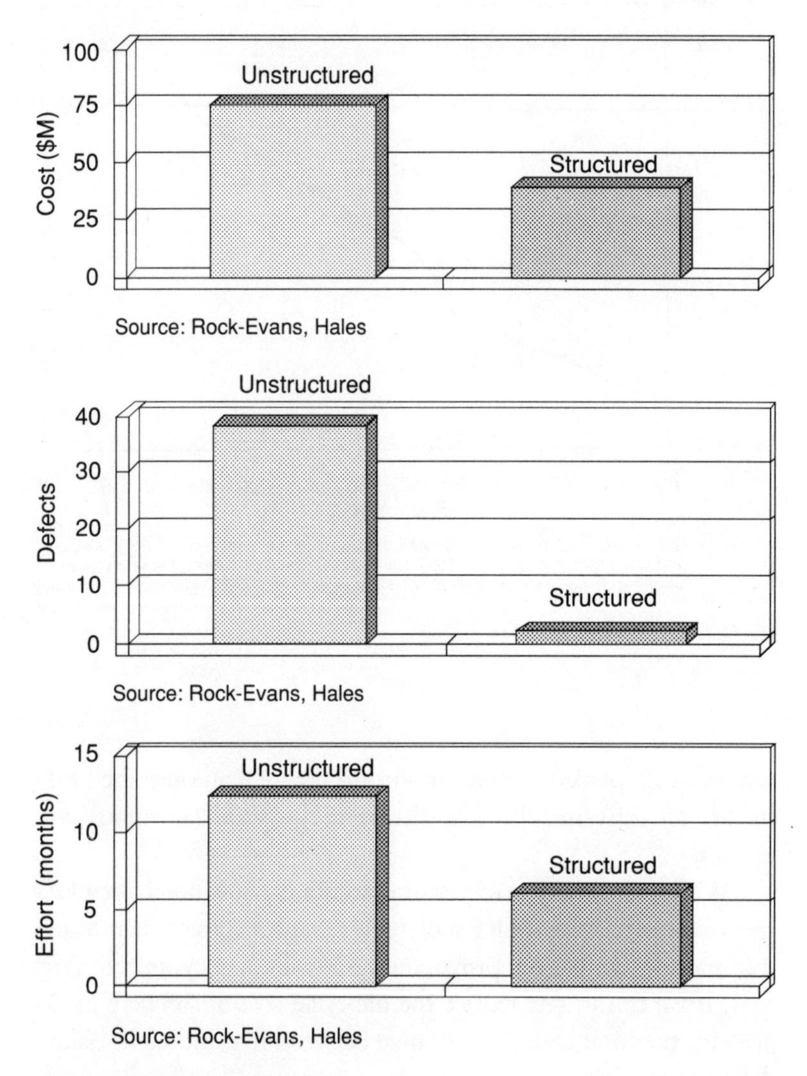

Figures 3-4A, B, and C. Structure versus cost/defects/time to change.

structure. It cost only half as much to modify the structured software and it took less than half the time. As an added benefit, there were about one-eighth the number of errors in the structured software.

Another indication of increases in productivity that may accrue from well-structured software is shown in figure 3-5. The points on the graph represent software size and productivity for development of some systems programmed in Ada. One of those systems, the Command Center Processing and Display System Replacement (CCPDS-R), was developed with special attention to designing a system architecture and tools that facilitate its modification. The original system was then significantly modified to produce two new versions. Productivity data for the two modified versions of the system are shown in boxes in figure 3-5. The high overall productivity is due in part to the architecture that accommodated these changes and in part to tools that facilitated making changes. Further benefits were realized because the architecture made general system services more accessible and, as a consequence, the application modifications were smaller than they might otherwise have been. The productivity data were adjusted for the reused and tool-generated software. This increased productivity is even more impressive when the usual negative relationship between productivity and system size is taken into account.

While important, the dollar cost of making changes to the system is only one concern; time is another. Operation Desert Storm

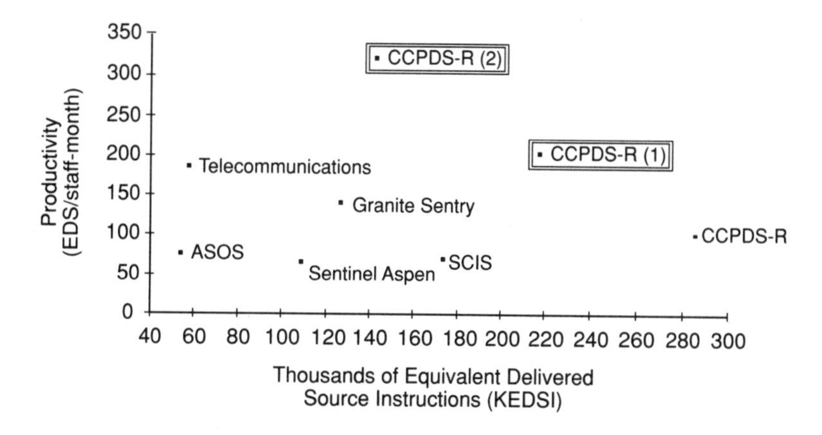

Figure 3-5. Maintenance productivity.

provided many examples of how well the flexibility of software served the allied cause, but there were also cases where we were not able to exploit software as we would have liked. Requests for changes to certain systems were made early in the campaign, but it was estimated that the desired changes would take 18 months. This was obviously unacceptable, and the military found it hard to understand why it should take so long, given that software is supposed to offer great flexibility.

Software does provide flexibility, but it must be designed from the start with an architecture that allows it to do so. Furthermore, everyone concerned must preserve the integrity of the architecture; otherwise, flexibility can be lost through the process of change. As an example, figures 3-6A and 3-6B are plots of the time it took to create each release of an IBM operating system and the number of modules affected in each release. The graphs show a progression; that is, it took longer and longer to modify the system as the system grew older. This was due to the growing complexity of the system — more and more modules had to be changed for each new release. The software

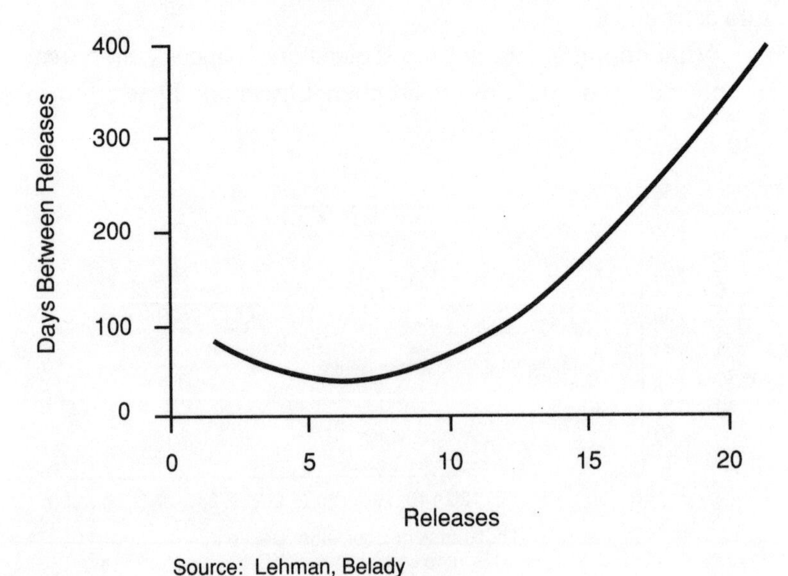

Figure 3-6A. Release interval versus system age.

Rapid response for the urgent needs of Desert Shield and Desert Storm created effective displays for the ABCCCIII consoles, leveraging the effectiveness of the displayed data.

structure degenerated, which made it more difficult to determine which modules had to be repaired. In addition, the pattern of regression testing had to be more extensive, since so many parts of the system had been affected by the modifications.

This complexity and uncertainty translates into more time and money, and the process begins a vicious circle — modifying the system makes the next modification even more difficult, time-consuming, and expensive.

Architecture: The Invisible Component

The DoD does not usually buy architectures — it buys systems that meet explicit functional and performance requirements specified by the user or the acquisition agent. In most cases, the DoD does not ask for an architecture to be delivered; it should therefore come as no surprise that very few architectures are delivered.

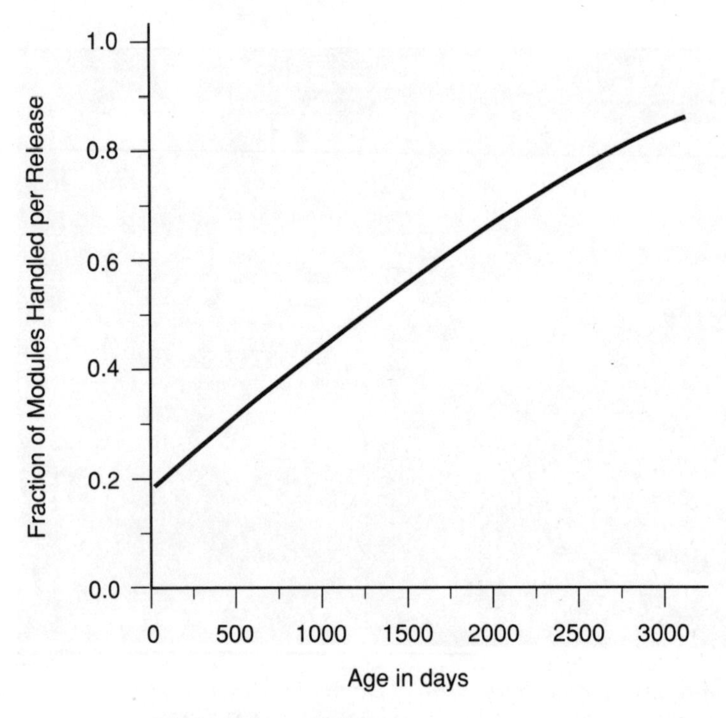

Figure 3-6B. Increasing complexity with age.

This is not to say that the DoD does not receive system architectures. Every system has some form of architecture, but the architecture the DoD receives may be quite convoluted and inflexible by the time the system moves from concept to fieldable implementation. There are no explicit specifications for the characteristics of an architecture, no formal tests of its capabilities, and no formal control of its structure to prevent arbitrary changes once it has been defined. This is one reason why architecture is fundamentally invisible — operational users are often not aware that an architecture is even present if it does not directly affect the functional capabilities they are using.

Yet architecture is the main determinant of a system's characteristics. The efficiency of the system, and thus its performance, depend on how the architecture handles the use of resources; architecture determines how the system sustains operations when parts of the system fail. The architecture also determines how maintainable the system is, *i.e.,* (1) how much effort is required to find and fix errors, (2) how easy it is to add new capabilities through software, and (3) how much is required to move the software to different computer hardware. Although they may be invisible to the user, these characteristics, which are all determined by architecture, are very visible

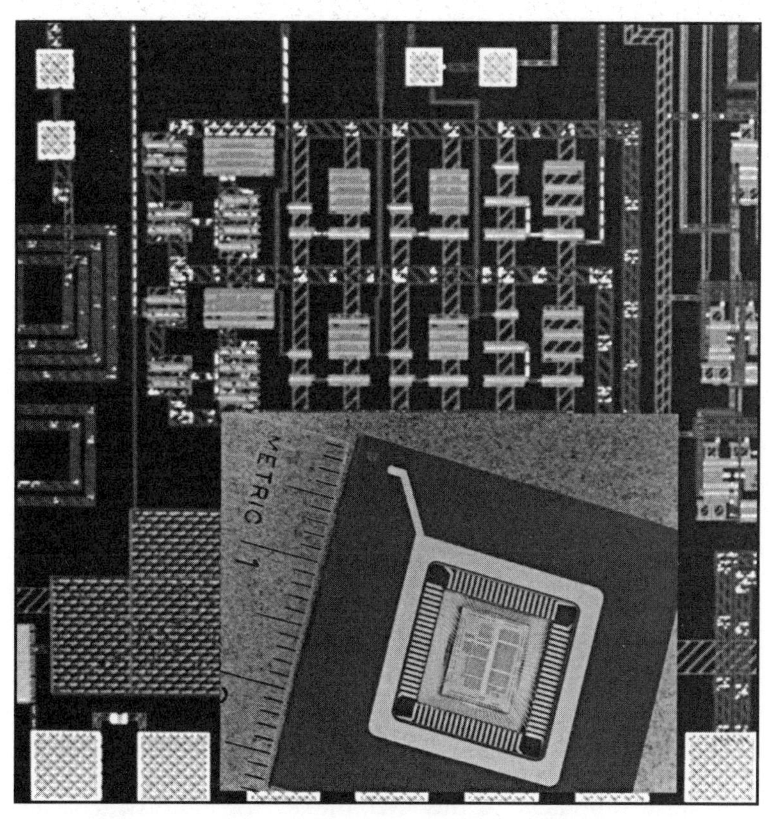

Powerful signal-processing functions can now be embedded in complex systems, greatly simplifying their architecture. The MITRE board and chip shown here perform fast Fourier transforms and electronic-to-photonic conversion.

to developers and maintainers who must modify and add to the operational capabilities of the system.

If the DoD wants to buy architectures, it will first have to know how to ask for them, specify them, test them, demonstrate them, and prevent them from degenerating; in short, it will have to perform all the operations that it performs now when buying other products.

In addition, DoD must perform a new task that is currently not part of its acquisition strategy: It must maintain explicit control of the architecture for the life of the system. One way of accomplishing this is to add architecture to the other aspects of the system that are controlled by configuration management. Since the maintenance phase contains a large fraction of the system's software costs, the ultimate maintainer of the system — and thus, the government — must eventually assume control of the architecture. This will require a significant change in the way the government currently views architecture and its importance.

Architecture: A Definition

There is no single, commonly accepted definition of a digital system architecture. In the broadest sense, "architecture" is defined as a system or style of building having certain characteristics of structure. When applied to digital computer systems, architecture includes the hardware and software components, their interfaces, and the execution concept that underlies system processing.

The simplest level of a system architecture defines how the hardware and software that make up the system are partitioned into components, and how software components are assigned to hardware components. Figure 3-7 is an oversimplified example (only primary functions are shown) of a fighter aircraft's federated hardware and software structure, which consists of separate computers networked on a standard bus with individual software functions assigned to the individual computers. At this level, the defense industry generally does a fairly thorough job of understanding architecture, mainly because developers need to understand how much hardware of which types they need to buy.

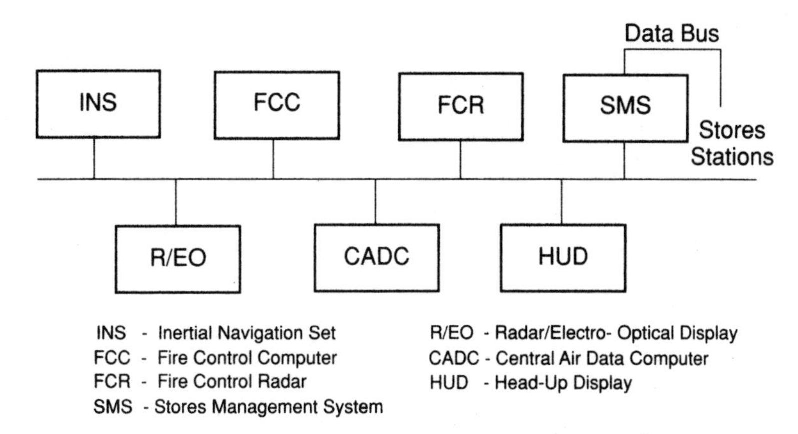

Figure 3-7. Hardware and software structure.

Figure 3-8 is another view of the digital system architecture for the same aircraft, showing both the application software in the previous figure and the software that performs system-wide functions. The functions can be described as grouped into layers; in this view, software in any layer may use software only in its own layer or the layer below it. The computers in the lowest layer represent the segregation of hardware from software to increase their independence and to enhance software portability. This is an example of the first part of the definition of architecture — the arrangement of hardware and software components (namely, the structure).

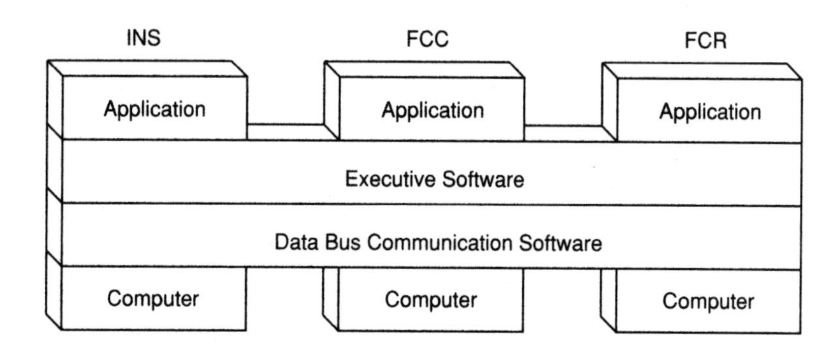

Figure 3-8. Digital system architecture.

The second element in the definition deals with the interfaces among key elements, for example, data communications according to a standard protocol (MIL-STD-1553). All computer-to-computer messages in the aircraft's avionics architecture must use this protocol; hence, adding new computers and new functions to the system is relatively simple (from a communications perspective) as long as the data bus has the needed capacity.

The third element in the definition of architecture is the execution concept. In the sample avionics system previously shown, this concept is based on the cyclic execution of each function, precisely timed to repeat the computation on a planned schedule. Neither of the two figures represent this dimension of the architecture. In fact, there is no one common notation for describing all aspects of an architecture.

Data-flow diagrams can show the execution concept of the architecture of a system (see figure 3-9). In this view, the sequence of processing, and which hardware and software components are involved as specific data move through a system, are apparent. This end-to-end view of the system's treatment of an external input is called a string; in actuality, there are many levels of detail that can be represented by a hierarchy of data flows for the same string. A string

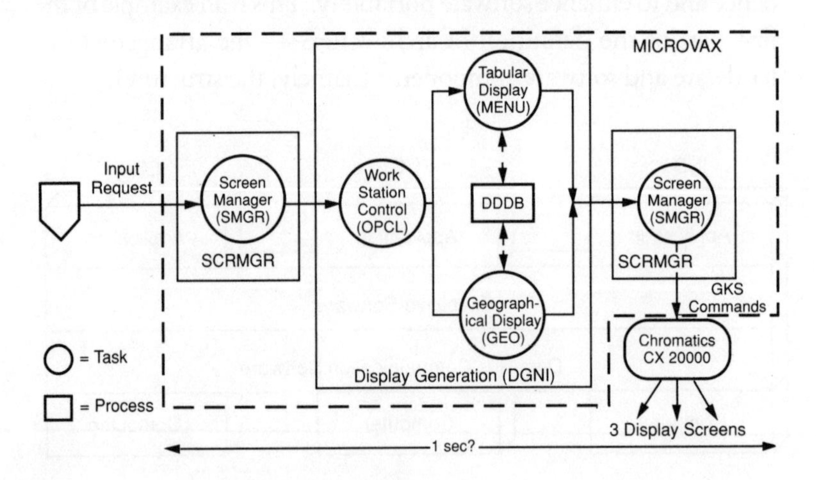

Figure 3-9. Data flow reference model.

is useful for assuring users that the system will perform the right functions on their data; it is also useful for estimating and controlling the time the system will take to respond to an input. What complicates the design of an architecture to meet response times is the large number of such strings that may be awaiting execution at the same time (as when many sensor reports are received or must be transmitted) and the contention over which string will use shared resources such as computers and communications lines.

To understand the timing aspects of a system, it is often necessary to develop a simulation that models the system architecture and the load on hardware and software components or to execute benchmark software on the actual hardware. The validity of the results depends on how accurately the load, the data flows, the hardware speed and capacity, and the timing of individual processes are represented in the model — even the most elaborate model yields useless results if the parameters are not accurate. The designer of the architecture must be given accurate information to design the architecture and to evaluate its performance; in other words, it is essential that there be good communications between modelers and architects or designers.

Since demands on the hardware resources will change over time, the architecture must provide the flexibility necessary to upgrade hardware to faster or larger processors so that requirements for increased processing loads or faster response time can be accommodated. Similar increases in bandwidth may be necessary in communications hardware to provide for increased loads. Models that correspond to an architecture can be useful in planning for and evaluating the effect of changes in the hardware configuration of a system architecture to meet new demands.

Taking structure, interfaces, and execution concept together produces one definition of architecture.

Of course, different vendors interpret the software part of the architecture in different ways. Figure 3-10 illustrates one vendor's view of software architecture. In many cases, commercial companies can provide off-the-shelf components for the general system capabilities of DoD systems; in addition, groups of commercial hardware and

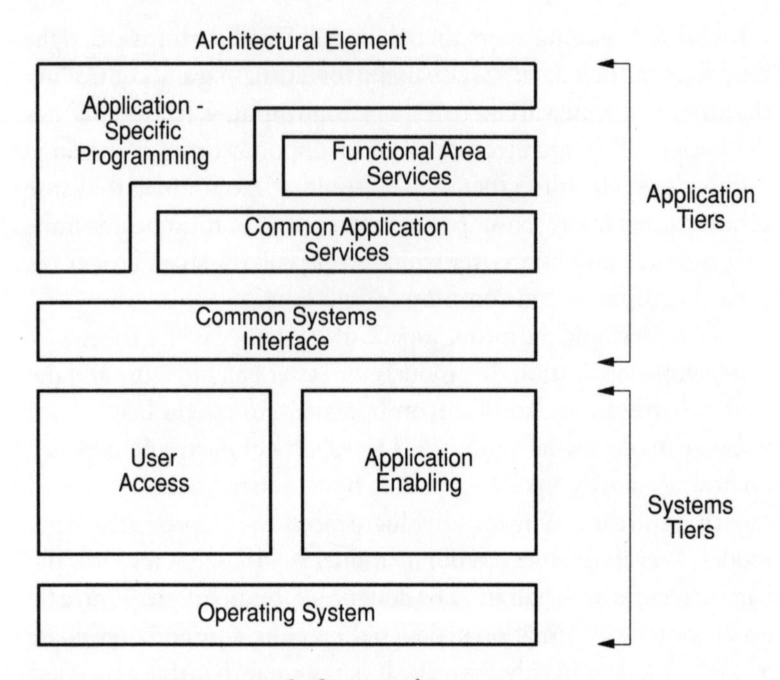

Figure 3-10. IBM view of software architecture.

software vendors are defining standard interfaces among layers and components. These "open" system architectures may provide the flexibility necessary to integrate, with a minimum of effort and system disruption, new hardware and software components having improved capability or maintainability. For example, the International Standards Organization (ISO) Open Systems Interconnection reference model defines the functions of each layer and the protocols for peer-level layers of a communications interface. Standards of this type permit the upgrading of elements of the system at particular layers without requiring the alteration of elements at other layers.

Figure 3-10 also illustrates the concept of service layers in the part of the architecture that is developed uniquely for one class of application (such as command centers or communications systems). These or other services must include error detection and recovery, interprocess communications, scheduling, and synchronization of

The availability of smaller, higher-capacity mass memory storage devices, such as the optical discs shown here, opens new architecture options.

processes. At this level of architecture, we must rely on the applications designers for standards within their design, as well as the quality control procedures to ensure adherence to their standards.

Architecture: Ramifications

The lack of a good architecture has a serious bearing on the cost, effectiveness, and availability of DoD systems. For many applications where high reliability and availability are necessary, the architectural concepts must incorporate failure management as well as other mission requirements. Trouble follows when this is not part of the initial architectural design.

Error handling is a critical component of any system, since errors will inevitably occur. Most systems have software to detect errors and to recover from an error when it is detected (for example, when a numerical value goes beyond expected bounds or when an operator pushes the wrong button). Given the critical nature of most DoD systems, it is crucial to keep the system in operation when errors occur. When we leave it to each programmer who has developed a part of a system to determine how to handle errors, the result is an unintegrated set of sometimes widely varying procedures that are often incompatible and even dangerous.

Recently, MITRE scanned the software for a large, safety-critical command and control system, and identified in the code more than 200 instances of incorrect error handling. In many cases, the system detected the errors and then either ignored them or passed them to another part of the system that could not handle them. What was missing was a consistent, coherent, system-wide error-handling strategy, a critical attribute of effective architecture. Furthermore, there was no method of ensuring that individual programmers adhered to the failure management standards that should have been established with the architecture.

The Complexity of Architecture

Perhaps the main reason that we fail to address these different aspects of system architecture lies in the increasingly complex nature

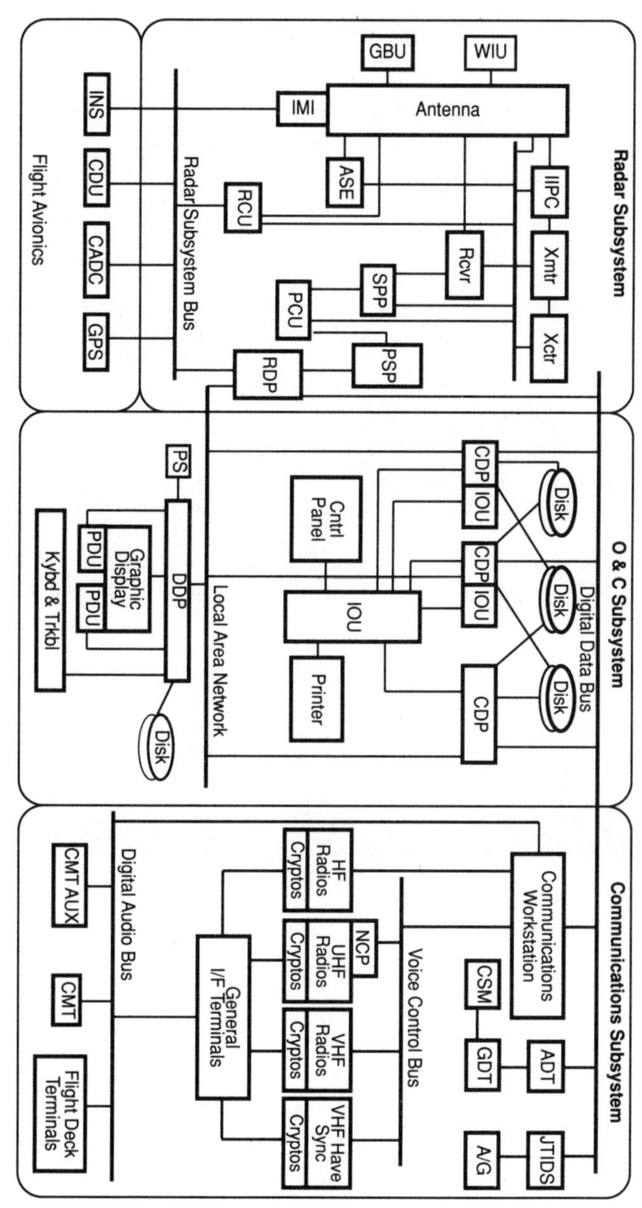

Figure 3-11. Joint STARS system architecture.

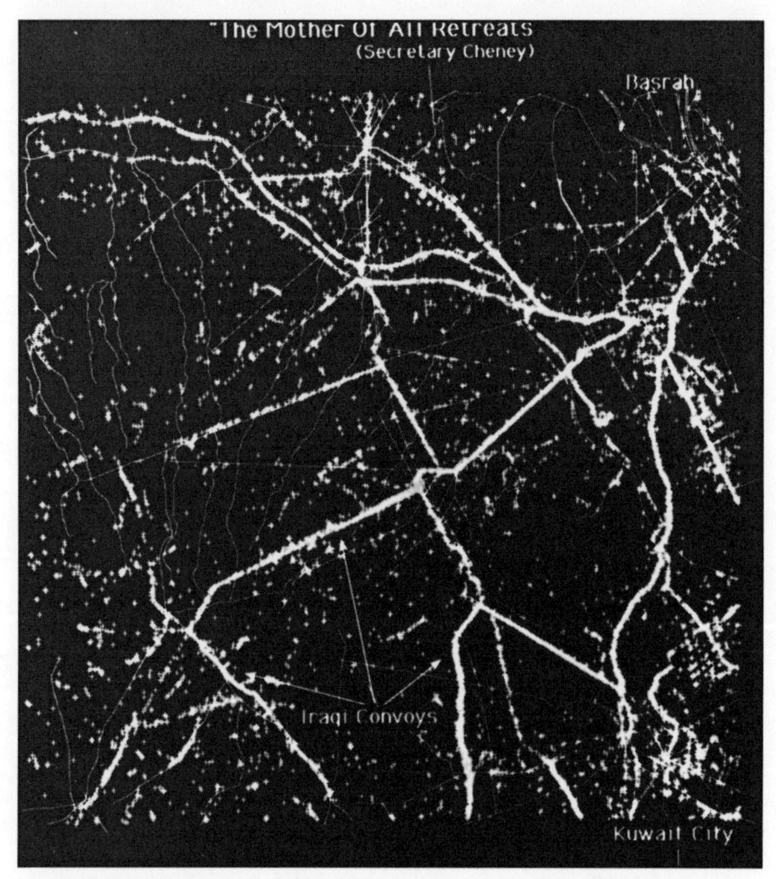

Joint STARS "Mother of All Retreats" photo shows Iraqi troops withdrawing from Kuwait by the thousands.

of the systems we build. Figure 3-11 illustrates the top-level system architecture of the Joint Surveillance Target Attack Radar System (Joint STARS). The actual architecture includes many more computers, many different data buses, and a large number of other components (not shown in the figure) to perform its demanding mission. The result is a system whose size and complexity make it difficult for developers to consider the many different aspects of architecture.

At the same time, the larger and more complicated the system, the more important good structure becomes. Developing and maintaining structure may be very difficult in a system of such complexity, but the rewards for doing so are even greater. These rewards include higher quality during the initial development, lower life-cycle software costs, and the increased likelihood that the system will remain in operation far longer (due to its greater flexibility and ease of upgrading). Furthermore, the reuse of known and expandable architectures will reduce the amount of new software that has to be developed and increase the quality of the systems that use them.

Architecture: The Waiting Solution

Technical Focus

At the start of a development program, when consideration of architecture pays the greatest dividends, the technical focus in the typical DoD program is often not on architecture. Rather,

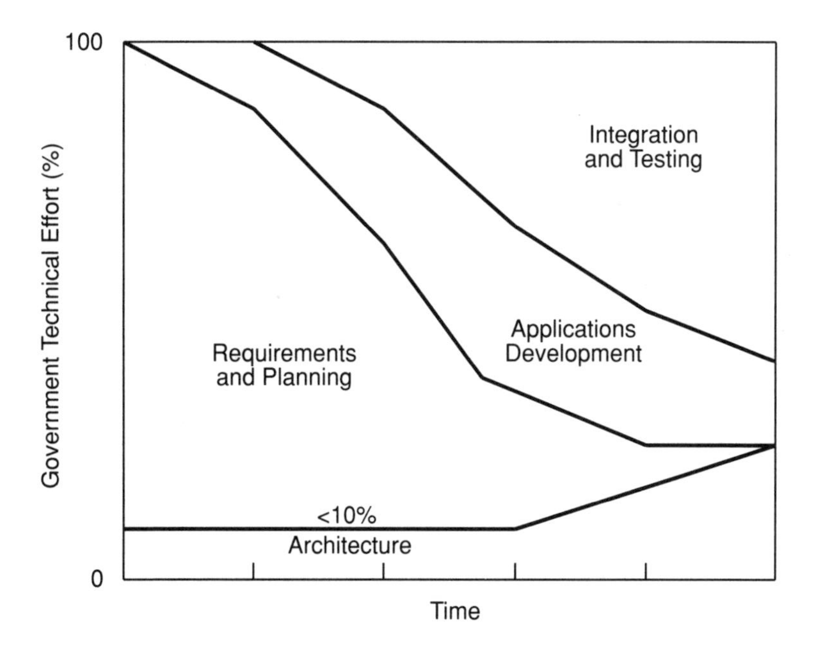

Figure 3-12. Technical focus (estimated).

functional and performance requirements are generally focused on by both DoD and the contractor (refer to figure 3-12).

This lack of attention to architecture occurs because the government expresses its requirements in terms of specific, measurable system functions and performance requirements that matter to the immediate user, and not in terms of flexibility, which matters to the maintainer and next-generation user. Government standards, such as DOD-STD-2167A, require proof that a design satisfies all functional requirements, not that it is adaptable to change. Design documentation and reviews track individual system and software components, with less attention on the overall architecture until the components are integrated.

As the figure shows, the failure to consider architecture throughout the program's development has serious ramifications as time goes on. The performance and control problems described earlier begin to mount, and the contractor is often forced to call on Red Teams and other desperate measures to modify the original architecture. Since it is done in haste and then only to allow the product to meet the specifications, this last-minute change in architecture does little to ensure the necessary efficiency and flexibility, and usually results in further degeneration of the basic design. We need to change the process.

Faulty Emphasis

Both government and industry typically put almost all their efforts into the initial performance and functionality of a program, despite the fact that these will change substantially over the life of the system. At the same time, there is a near-total lack of attention to an architectural baseline that would form a stable foundation for incorporating the system's changing requirements.

What we do ask for does not address the important architectural issues. For example, we state that the system shall be modular, but we don't demand a good way to partition it into modules that will allow future expansion and change.

We also specify requirements for system growth in an ineffective way that does not relate to operational capabilities, such as adding new message types or increasing message traffic. Timing and sizing margins — for example, half the time and twice the memory — cannot ensure that the resources provided are allocated in such a way that they can be used to meet future requirements.

With the advent of distributed systems, timing and communications bandwidth margins become important in providing for future growth. The government needs to ensure that growth is expressed in operational terms, and not just in physical terms.

Because of the government emphasis on meeting immediate requirements within schedule and cost, even industry perceives that the government is not seriously interested in controlling maintenance costs. In a 1990 Air Force Scientific Advisory Board study of software maintenance, 123 businesses were asked what the government thinks is important when awarding software contracts. The results are summarized in table 3-1.

Table 3-1 Numerical Ranking (Out of 7 Points) by Business

Overall project cost	6.2
Proposed product performance	5.5
Contractor experience in area	5.5
Timeliness	5.3
Last contract an advantage	4.8
Project software development cost	4.6
Contractor software capability	4.4
Ease of software maintenance	3.4
Software maintenance cost	3.3
Software portability	2.9

The Scientific Advisory Board view of the government's stress on cost and system performance, rather than long-term maintenance, is readily apparent.

Commercial Architecture Trends

Both commercial software users and commercial software vendors have become more concerned with architecture. The software buyer wishes to preserve the hardware and software investment without giving up the ability to upgrade portions of the system over time. The software vendor wishes to compete for a broader share of the market by making the components operate on more hardware platforms and interoperate with a larger selection of other software components. This has led to the banding together of commercial vendors to define open system architectures with interface standards among the components. Such architectures, like that in figure 3-10, provide commonly used services and interfaces for using them in building specific applications.

These commercial architecture trends can do nothing but help DoD software efforts, because DoD is a large buyer of software and hardware that support these interoperability standards. Even embedded, special-purpose militarized systems rely heavily on commercial systems to assist in software support. The DoD cannot try to take the lead because these standards are driven by the commercial marketplace; however, the DoD can use to advantage the opportunities in the commercial market for open architecture standards. Unfortunately, these commercial standards cannot include the service standards that are heavily application-dependent; these must be left to the application designer to establish and implement.

Reference Models

The government has begun to use the growing body of commercial standards by defining what has become known as a reference model for a set of closely related systems. A reference model specifies a common set of components in terms of the services each component provides, the interfaces for accessing those services, and the grouping of components into layers. For most of the common services such as user interface, database management, and communications, there are commercial standards defined. Figure 3-13 is an example of a reference model defined by the National Institute of Standards and

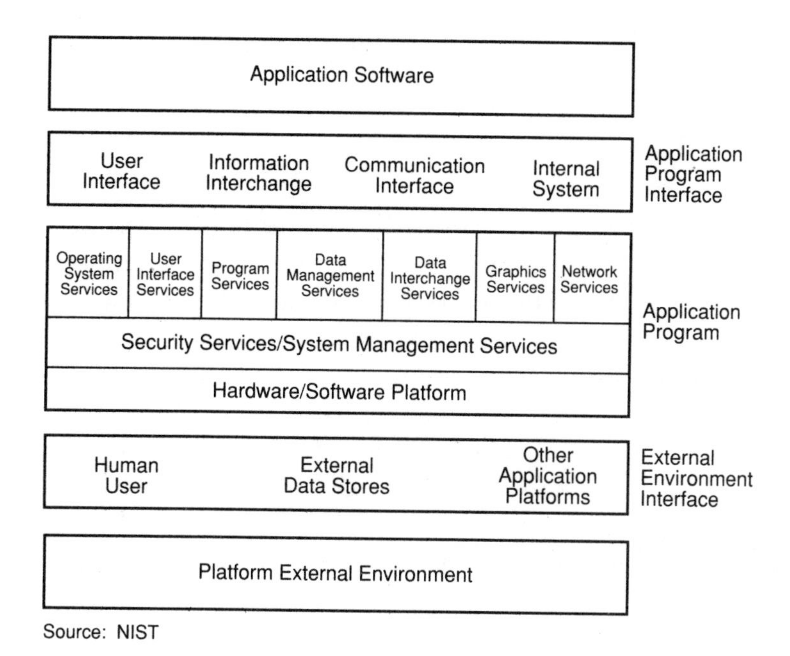

Source: NIST

Figure 3-13. NIST open system environment reference model.

Technology (NIST) for DoD information systems to provide a set of functions with interoperability, portability, and scalability across a network of heterogeneous hardware and software platforms. Generic reference models, such as that of NIST, can be adopted for a specific group of applications and made more detailed as shown in Figure 3-14, which is the reference model for the DoD Intelligence Information System (DODIIS). In this case, the allowable standards are named and the components arranged in layers that define the scope of their interaction.

The use of reference models is a positive step in providing a higher level of conformance among families of systems, and it can foster greater reuse of software. It does not, however, address important design decisions that must be made in selecting and combining specific components, in the detailed design of a system's dynamic behavior to meet its requirements, and in ensuring conformance of the architecture to the specific implementation of the reference model throughout the system's life.

Availability of Tools

The commercial market is also in the lead in providing tools that support the designer in generating and documenting architectures. There are tools to link software components together and assign them to processors. Other tools enable developers to analyze the linkage between different software modules, the control flow, the flow of data, and the timing of the various operations, and to assess and improve architectures.

Many tools can only perform their analyses after the software is already written. These tools can still be used to understand what has been developed and to evaluate how easily it can be modified, before it is fielded or later. The investment may be small, and the potential payoff large. Table 3-2 lists some representative examples of available tools.

Table 3-2 Representative Software Tools

Name	Vendor	Analyzes
Logiscope	Verilog	Module structure, path coverage
ACT/BAT	McCabe	Flow graphs, structure graphs
ADAS	CADRE	Dynamic behavior, timing
STATEMATE	i-Logix	Structure, dynamic behavior
CPN	Meta	Dynamic behavior, simulation
Adagen	MarkV	Ada static structure, dynamic behavior
CARDtools	Ready	Timing threads
TAGS	Teledyne Brown	Dynamic behavior, simulation

The government must acquire such tools and use them if it is going to buy architectures and understand them.

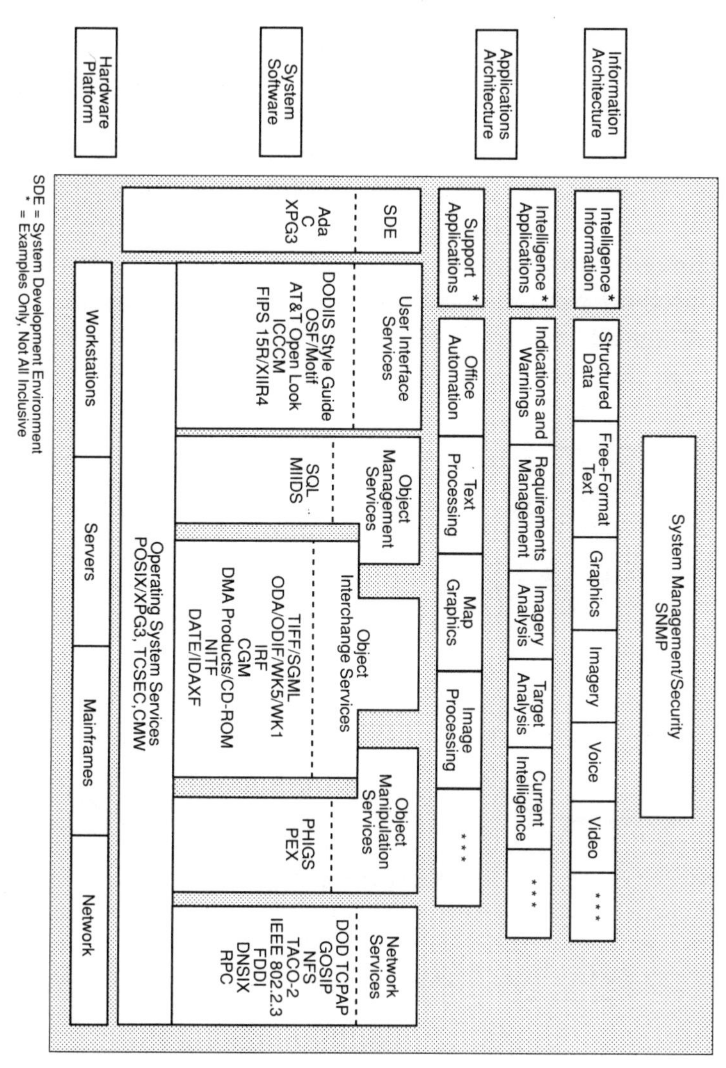

Figure 3-14. DODIIS reference model.

In addition to commercially available tools, project-specific tools can improve the productivity of software development for a specific architectural design. Referring to the CCPDS-R program again, the contractor developed a tool to generate automatically the communications software that linked applications. The applications programmer needed only to list the elements of data that were required from each application and were necessary to each application. The tool used this information to generate communications following a standard pattern that were efficient and correct.

Recommendations: Buying Architectures

Good architecture can potentially provide significant cost savings as well as greatly increased system flexibility. To obtain these benefits, we must put architectural requirements in system specifications,

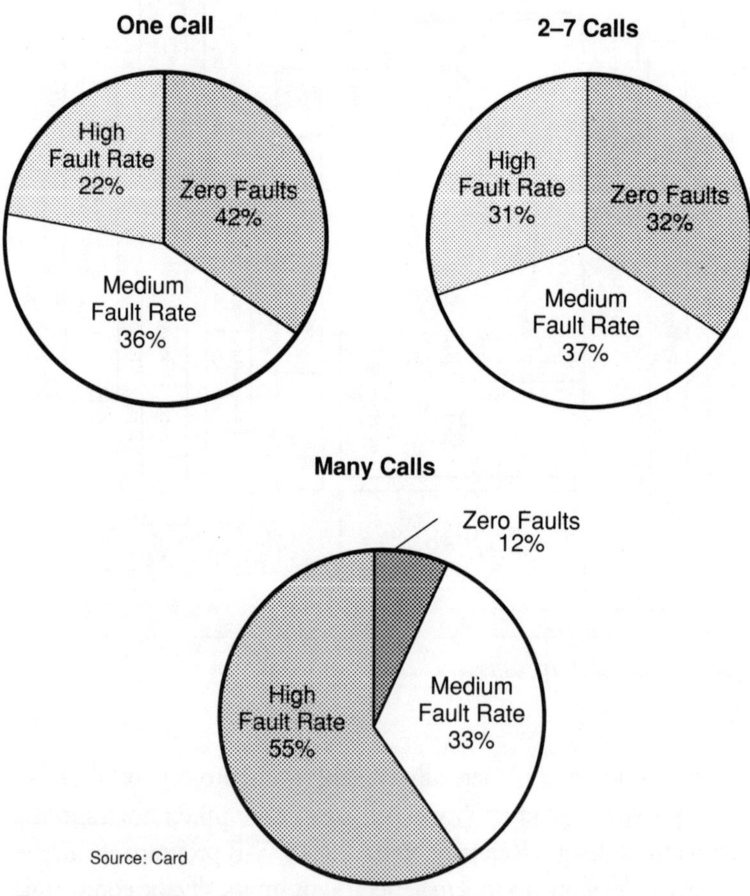

Figure 3-15. Effect of control structure on errors.

emphasize the early satisfaction of these architectural requirements, give contractors incentives to use proven architectural concepts, and control the architectural configuration over the life cycle of the system. We believe this can be done.

System Specification

Since contract requirements drive the entire development of a system, the surest way to ensure adherence to a sound architecture is to put architectural requirements in the system specification. This does not mean that the specification will define the exact architecture to be used, but rather that it will specify what the architecture is to do. In cases when the application domain is well understood and a sound architecture is already available, the government may find it in its best interest to be more restrictive than in other situations lacking such a clear precedent.

To specify accurately the criteria that architectures must meet, we must also determine how to qualify them. There are few measures of system designs that accurately predict flexibility and expandability. We will have to depend on a combination of techniques, including demonstrations that the system can be modified as well as analyses of features of the architecture. We are beginning to establish relationships between measurable features and the rate of errors as well as ease of change. For example, the more calls a module makes on other modules, the more errors occur, as figure 3-15 shows.

Early Satisfaction of Architectural Requirements

To reap the maximum possible benefit from architectural requirements, we should specify that contractors cannot write large amounts of applications software until they have developed an architecture evaluated and approved by the DoD. The only applications software that should be written before this point is that necessary to help evaluate the architecture and reduce other serious risks, not to perform the actual task at hand. We can no longer afford the risk of developing architecture and applications concurrently; on the other hand, if contractors have successful architectures and control procedures that they have used before, they can use them again. In fact, the quality and effectiveness of a previous architecture as well as the tools available to support development of applications within the architecture should be an important factor in the selection of contractors on a program.

We should also control these architectures after we evaluate and approve them. Changes would be weighed against the need for future flexibility throughout the life of the program.

This prototype digital imaging network system for the U.S. Army Medical Research and Development Command was based on the concept of open system architecture.

Contractor Incentives

Contractors will have to be given incentives to change from their current emphasis on meeting immediate requirements to a longer-term view. They will have to set up their own controls to keep applications software writers from corrupting the architecture; in other words, during development, contractors will have the architecture under configuration control. Rules and standards have to be defined as part of the architecture. Tools should facilitate the integration and modification of components within the architecture so we know that the standards of the architecture are observed.

Contractors who have good architectural awareness should be treated better than those who do not. The development community needs to start working on architecture with the software maintainers to ensure that we deliver to them whatever is necessary for them to sustain and use the architecture.

Getting Started

The best approach to implementing these recommendations is for the DoD to put together a government and industry team to develop the specification and contractual language for buying architectures and the criteria for evaluating them. Such a team can use its experiences to determine what approaches have been successful in acquiring good architectures and what has not worked. We have already begun to achieve encouraging results from a partnership between a team consisting of members from the Electronic Systems Center of the Air Force and MITRE, and an industry team with representatives from six major defense system contractors.

Preliminary Results

The joint team is refining an initial concept for a process that leads to the acquisition of good architectures through the following steps:

(1) The government specifies the current requirements for the architecture and adds a "life cycle vision" of its plans for future technology insertion, growth in the capacity and response time performance of the system, and likely changes in functional requirements;

(2) Industry responds with technical proposals describing the top-level system architecture and how it will meet this vision, and the process and tools by which the architecture will be developed, documented, evaluated, and monitored;
(3) The selected contractor and the government collaborate on the incremental detailed design where the government's role is to provide more detail on its vision while the developer continues to design to current requirements and the future vision;
(4) During the detailed design process, there is continuous assessment of the design by simulation, analysis, and execution, to affirm that the design assumptions are correct, to demonstrate that the design meets requirements, and to estimate the effort that would be required to meet the vision; and
(5) Once the architecture is agreed upon, periodic checks are made to see that the developer has adhered to the proposed development process and to the baseline architecture and that the architecture continues to meet its requirements. As part of this process, the developer produces documentation for the maintainer of the architecture.

The joint team has also proposed that a good architecture should incorporate the concept of a system manager, a set of software that controls the connections for basic system services and for all interfaces among major functional components. Functional components would not directly interact with each other. This would allow new components to be added with minimum disruption to existing components. Components could be developed in parallel and integrated through the central controller. Missing components could be simulated to help in the testing of components.

Further Work

More work will be done to arrive at a set of information that conveys the important aspects of an architecture and the criteria and methods for evaluating an architecture. We need to formulate the design patterns that are successful as well as those that cause problems. For the many existing systems that must be maintained and upgraded, we must work to introduce architectural improvements without disrupting operational use of the system and without unnecessary replacement of software. In the long term, research on methods of generating systems from higher-level specifications and from existing components will contribute to software systems that can be changed more easily and more rapidly. The insights of the current joint government-industry team on architecture will be of considerable benefit to the DoD during this time of changing missions, shrinking budgets, and increased need for flexible systems.

Chapter 4
Development Of DoD Software-Intensive Systems Using Commercial Off-the-Shelf Products

The use of COTS products offers DoD cost-effectiveness and flexibility, but must be approached with caution.

The use of commercial off-the-shelf (COTS) products in the development of systems for the Department of Defense is crucial today. Significant reduction in the current DoD budget and the expectation of further reductions in the near future will force the defense community to build lower-cost systems that do not depend heavily on customized hardware and software. The concept of "open system architecture," where individual subsystems designed and manufactured by different industrial organizations can (by virtue of common interface standards) be integrated into systems of high performance, is making its appearance in the commercial marketplace. This creates the possibility that large-scale military systems can be integrated at low cost from commercial products, with relatively short development schedules. The DoD will need to exploit this possibility.

An additional impetus toward the use of commercial products is the public's (and various military commanders') growing familiarity with products that are readily available on the market for home and business use. As part of their personal lives, military decision-makers are becoming aware of the powerful computing and software options that can be obtained off the shelf, and will be unwilling to believe that custom development is necessary to achieve desired performance. As a result, commercial systems will become the norm for military systems, with custom-designed equipment as the exception.

Over the last few years, substantial effort has been spent in attempts to integrate commercial products into military systems. From these efforts, the defense community has evolved a style and approach to dealing with COTS integration. Rapid prototyping is now being employed extensively as part of the design process. Many professionals in the prototype arena are knowledgeable about open system standards; this was not true as recently as four or five years ago. Much more attention is being given to commercial standards, because they are evolving to be useful and because organizations in the commercial world have been formed to deal with them.

Participants and Their Roles

There are three major participants in a typical military development: (1) the Using Command, or User, who sets the system requirements and who has final responsibility for operating the production system, (2) the Development Command, or Developer, who is given responsibility for developing a system meeting the User's requirements, and (3) Industry, which actually designs, produces, and tests both the development versions and the final production versions of the desired system.

However, when the developments are software-intensive systems with an emphasis on the use of COTS, these typical roles are not always followed. Table 4-1 presents a list of projects for which the

Electronic Systems Center (ESC) of the Air Force was the development command, supported by the MITRE Corporation. The development approach, and the roles apportioned to the three participants, was somewhat different in each case.

Table 4-1 ESC Programs Showing a Variety of Participant Roles

System	User*	User's Role	Developer's Role	Industry's Role
Sentinel Byte	USAFE, PACAF, SAC, TAC, MAC	Generated requirements.	Built two COTS prototypes in stalled at AF sites. Success led to decision to buy 15 more.	Replicated developer's prototypes.
Granite Sentry	NORAD	Hired and managed support contractors.	Generated and analyzed system architecture, maximizing COTS.	Provided software support to user.
NORTIC	NORAD	Generated requirements.	Developed system architecture, system design. Selected a contractor to buy COTS components. Selected another support contractor to integrate, test, and maintain system.	Bought COTS components to implement developer's architecture. Integrated, tested, and maintained system.

Table 4-1, *continued*

System	User*	User's Role	Developer's Role	Industry's Role
Message Handling System	DIA	Generated requirements.	Generated specification; hired contractor to demonstrate initial prototype.	Designed and built prototype, maximizing COTS.
ABCCC	TAC [ACC]	Generated requirements.	Generated product specification. Selected two contractors to build prototypes.	Designed and built prototypes and production version.
Mission Planning System	TAC [ACC]	Built 3/4 of system, asked developer to complete.	Transitioned system to more open architecture, so that capability could grow by addition of COTS products.	Designed and built system, using developer's architecture.

*Note on Users: SAC and TAC deactivated June 1, 1992. Now replaced by Air Combat Command (ACC). Air Mobility Command (AMC) replaced MAC.

Judgment

The problem in using commercial products is relating the commercially available equipment and software to the mission need. This problem involves the following different skills and judgment, as shown in table 4-2.

Table 4-2 Kinds of Judgment Required for Various Issues

Issue	Kind of Judgment
What products are available on the market?	Oversight of the whole commercial market.
Can commercial products be integrated, even if individual products have adequate capability?	Technical judgment
Will the completed COTS system be operationally useful?	User judgment
How much will it cost to add capability at some later time?	Technical judgment
How difficult will the COTS system be to modify?	Technical judgment
To what extent is the user willing to compromise on the system requirements, balancing what is desired with what is readily attainable?	Mission judgment

Two Important Philosophical Issues

In each of the cases described above, different judgments were made about the mixture of roles that, when coupled with business decisions, led to a different acquisition style.

While we agree that there is no one best acquisition style, the different styles used in the examples above are not a sign of healthy variety. They did not result from a set of logical decisions, but instead were driven by the varying political and economic environment surrounding each development. No clear method or policy was available to aid in the selection of acquisition approaches, because the defense community has not yet dealt adequately with some important philosophical issues regarding developments incorporating commercial products. Two of these issues are discussed below.

Performance specifications

The topic of performance specifications has not yet been resolved satisfactorily. In Air Force system development, we have traditionally begun by first writing performance and functional specifications. This is typically how the government explains what it wants a contractor to do, thereby providing the contractor with a firm basis upon which to accept the risk of building the system at a fixed cost.

However, when a system is to be developed using commercial products, the design lattitude normally available when building a system composed of custom components is severely constrained. Design freedom is implicit in the idea of a specification. For programs driven by the desire to integrate unmodified commercial products, freedom of design is a false assumption; there is, in fact, very little lattitude in design. The concept of writing a specification does not jibe with the idea of building systems from commercial parts.

In principle, there need be no conflict between specifications and the use of COTS products: Before making a bid, industry could perform the tradeoff studies needed to identify the applicable COTS products, select those that represent the best match to the specification, and verify that the selected products could actually be integrated within the proposed schedule and cost. However, projects based on commercial products virtually always have small budgets. It is unreasonable to expect a company to invest in extensive tradeoff studies before it bids on a project that offers only modest returns.

On the other hand, Chapter 3's discussion on architecture assumed that industry would be responsible for the selection of a suitable system architecture and its COTS components. One way to resolve this economic impasse is to choose architectures that can survive intact even if some of the COTS components originally picked by industry prove to be inadequate and need to be replaced. Another way (assumed in this chapter) is to give the design responsibility for the architecture to the government.

In practice, industry must use its current (relatively limited) knowledge about commercial components and make a calculated guess about whether those specific products can be integrated to meet

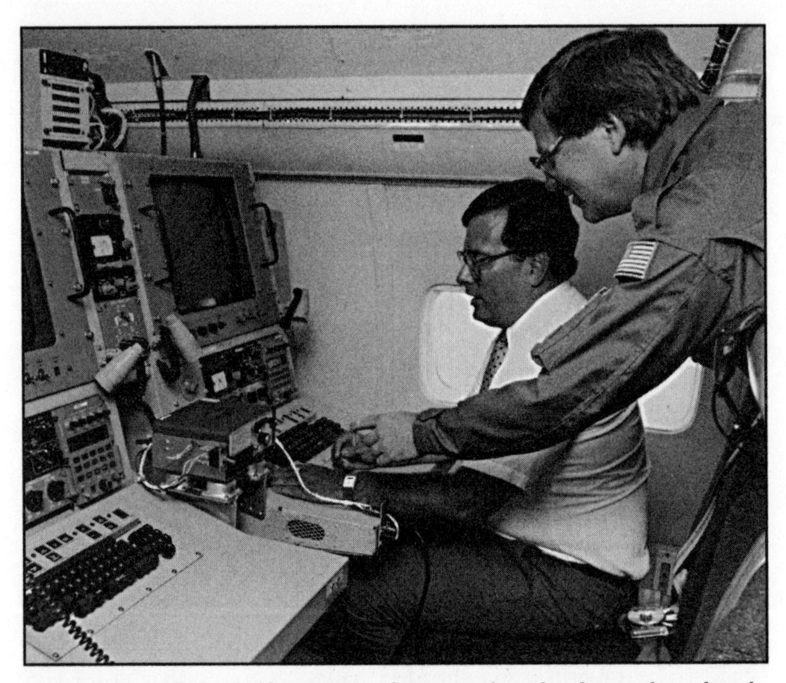

Commercial off-the-shelf (COTS) software package has been adapted and modified for DoD use in the on-board display systems for the AWACS and Joint STARS, with a savings of time and money.

the specifications. If the guess is wrong, there are no options available for recovery, because typically the government has budgeted no money to do any custom design. For example, some development projects at ESC have been cancelled because the efforts were started with a specification and were later discovered to be infeasible using the selected COTS products; both industry and the government took large losses. A way must be found to remove specifications from the process of buying COTS-based systems.

System architecture

A second consideration not adequately resolved is system architecture, which bears on the whole issue of life-cycle costs in software-intensive systems. It is known that a large percentage of a software system changes and evolves over its life cycle, and that a large portion of the cost of owning a software system occurs after the initial development is finished and the software is installed. NASA estimates that between 60 and 80 percent of the cost of software is incurred after it is shipped to the user. Of that fraction, about 80 percent is spent adding new features, because either the original capability was unsatisfactory or the mission changed. Similarly, at ESC about half the software effort is spent reworking what was initially developed, even during a development. In total, about 75 percent of C^3I software development efforts are spent responding to changes in the original specifications. The nature of C^3I development precludes firm adherence to specifications.

To accommodate to such variability in requirements, we can insist (as discussed in Chapter 3) on having a system architecture that is designed to make applications readily changeable. There seems to be no point in specifying every system function in exact detail, because most functions will undoubtedly be changed somewhat during the course of development. Without an architecture that supports the ability to change, we will be forced to spend substantial development funds changing functions that have been optimized to the wrong goals.

This situation arises quite naturally, because the Using Commands believe they know exactly what is needed at the beginning of the development process. The Development Commands believe their job is to sell what the users defines as their requirements. Thus the development process does not promulgate any scheme for thinking about adjustments. In fact, the current process discourages thinking about adjustments, because success is defined as meeting the specification.

Reducing the Dependence on Specifications

How can the problem of changing specifications be addressed? A new process must be created, beginning with a team whose only job is to devise an architecture that will support making the system changeable. The team should develop an architecture that readily allows for change of functions. If they have a sound architectural basis, they will know that, even if they make a large number of errors building the functions, they will have created a scheme that makes it inexpensive to make changes. Today, most of the effort goes into looking at the displays to be sure that they are right, and examining the exact functions to be sure they are implemented correctly, rather than verifying that the architecture is adjustable.

The government only recently has initiated C^3I development efforts by specifying a system architecture explicitly designed for changeability, and to my knowledge has never asked that a given architecture be tested for its ability to support change. No method exists today for generating such a specification, and system engineers have no satisfactory way to evaluate architectures. The joint government-industry team mentioned in Chapter 3 is striving to develop the necessary techniques, standards, and processes. Their success (or the success of other similar efforts) will be essential to the effective use of COTS in large defense systems.

A parallel may be drawn with the housing industry. Building houses is a very mature technology, with two major aspects: (1) the cosmetics of the house interior, concerning details such as the number of baths and the dimensions of the kitchen, and (2) the basic architectural design of the house.

These two aspects of house-building, cosmetics and architecture, are handled by two quite different sets of professionals. Inspectors are employed to make certain that state and local building codes are satisfied. Even though prospective home-buyers may not have technical knowledge about the spacing of studs or thickness of concrete, they are aware that it is prudent to have an expert check thoroughly these and many other parameters of the building; few people would purchase a house that had not undergone a careful review by a building-codes expert.

Confident that the house's basic architecture is sound and therefore extendable and changeable, the consumers can specify the details of internal functions, with the knowledge that those functions are likely to change as their economic circumstances change. They might, for example, put in bigger bathrooms or a larger kitchen.

In software systems, by analogy, the detailed functions inside the "house" are emphasized while the foundation of the "house" is virtually ignored. Although the users are fully knowledgeable about the desired functions of the software system, they have little skill in the "building codes" for software. Unfortunately, unlike the housing consumer, the software user as yet has no orientation toward bringing in the equivalent of the building-codes inspector, someone who can insist to the developer that the basic software architecture be sound and be readily able to accommodate change in the detailed functions.

Performance specifications create a negative environment for building systems from commercial products and for looking at the architecture from the viewpoint of changeability. How might complex C^3I software systems be developed from commercial products using a process that puts more emphasis on software architecture and is not based on performance specifications? The approach described below forms a team consisting of the users, the developers, and industry in various combinations, depending upon the development phase.

Understand the users' needs

The team, which could consist of the users, the developers, or a mixture, must understand the users' desires in detail.

Understand COTS products

The team must have detailed knowledge of as many available commercial products as possible. There are countless commercial products presently on the market, and new ones are appearing at a tremendous pace. A substantial effort is needed to understand available products.

It is not practical to set up an overhead activity in which a special team would become expert in all the commercial products; the team would be overwhelmed by the volume. The only feasible method is to form one or more teams that work on and oversee many different implementations of C^3 systems. In that way, knowledge would be gained about not only what products are available, but also how well they integrate and what problems occur with particular integrations. This knowledge would not be comprehensive, but over time would span a large fraction of the necessary ingredients in C^3 design.

Learn how to evaluate system architecture

Satisfactory methods for evaluating architectures must be developed. Two relatively crude approaches are occasionally employed now. The first involves building prototypes. According to my experience, the general interest in prototypes tends to be oriented toward display, analogous to buyers focussing on what color of bathroom wallpaper they want in their house. The second method is to evaluate prototypes with respect to their architecture. The second method is more important, indeed crucial, but this fact is not yet apparent to the typical software user. As a result, there are no funded efforts devoted to evaluating the architecture of prototypes; unfortunately, funding that could be used for this purpose is being expended on overhauling systems that have not been developed properly.

An important aspect of architecture evaluation is a method for rejecting products that seem attractive but which cannot be extended or expanded within the chosen architecture. There is a strong tendency to incorporate products if they appear to offer some attractive new feature, without careful examination to see whether the product will allow functional changes at a later date, and without an estimate of the cost required to add on the next new commercial capability. Although prototypes can help in estimating the cost of future changes, they are seldom used in this fashion, even though the cost issue is crucial.

COTS-based C^3 developments tend to be low-cost projects, driven by time and budget. If the development cost and the user's budget were known at the outset, an informed judgment could be made about what functional capabilities are affordable (whereas a specification must be generated at a time when the feasibility of using COTS products is in doubt, as well as the practicality of future additions). A better process would be to build a prototype capability, learn about cost and architecture from the prototyping exercise, discuss functional performance options and their associated cost with the users, and then, given a budget, decide what is possible.

Bring in industry differently

When the time arrives to bring industry into the team, a new method must be found to make the selection from among the bidders for a COTS C^3 development. Generally, the company selected by the government is the least expensive and most readily available one that is qualified, because both funds and schedule are severely constrained. To continue the house-building analogy, it is clear that we would not want to choose the contractor for our own home solely on the basis of cost and availability.

Having chosen the basic COTS elements of the architecture, a rational approach would favor companies that know the most about that class of COTS products, because it would be reasonable to expect those companies to be more successful at integrating the products at

The Multiple Launch Rocket System (MLRS) can launch 12 rockets in excess of 30 km in less than a minute. It is a tactical weapons system that uses COTS software. Courtesy of DoD.

the lowest cost. In a given development, if one company were found to be most knowledgeable about all the applicable COTS products, then it would seem appropriate to select that company. On the other hand, following this reasoning, if the selected architecture called for the integration of COTS elements A, B, C, and D, and if one company was found to be expert in integrating items A and B while a second company was expert in integrating items C and D, the government might choose *both* contractors. Unfortunately, the government would be fearful to take this latter course, because in effect it would be acting as the prime contractor responsible for integrating the system and would therefore be accountable for successful completion of the overall development.

However, the problem of government accountability may not be severe. The government's attitude toward accountability stems from custom-development projects that typically cost hundreds of millions of dollars; the fear of becoming entangled in expensive litigation in such cases is understandable. By contrast, the scale of COTS developments is typically in the range of four or five million dollars. When the government team has skills and knowledge to do the integration job more professionally than an industrial team (who may not know the user and product selection as well), the government should strongly consider taking on the integrator role.

Present Methods are Harmful to Industry

I believe that our current practices in the development of C^3 COTS-based systems are harmful to industry. A typical scenario illustrates the point: The Development Command requests proposals for a C^3 system built primarily from COTS products. This request is made even though there is no proof that the specifications can be met with COTS products. (Occasionally, feasibility is asserted by the developer because one COTS version of the desired system has been prototyped; however, the selected contractor is free to choose any set of COTS products it wishes. DoD does not know whether the *contractor's* selection of products can do the job — and neither does the contractor, for the reason stated earlier: No contractor can afford to invest in trying to prove feasibility before the bid is made.) After the selection, the contractor begins the development job, with the understanding that the specification must be met. When the contractor cannot meet the specifications within schedule, the monetary loss begins; and because these jobs are tightly budgeted, it takes only a few months of extra time before the contractor's profits vanish. The results of extended court contests are long delays, increased costs, and undesirable products. Industry bears the risk of meeting the users' requirements with its own selection of off-the-shelf products.

High-Risk and Low-Risk Phases of Development

Once the tradeoffs of meeting user desires versus the performance of COTS products have been made and are thoroughly understood, developing a COTS-based C^3 system becomes a low-risk project. The tradeoffs allow accurate estimation of integration costs and of the cost to add capability to the system incrementally. When the user's real budget is known, and when the user is willing to temper his or her requirements to stay within that budget, the risk becomes small. Generally, very mature COTS software is used, so the risk of encountering severe software bugs is relatively low.

Before the tradeoffs have been completed, however, the development risk is very high, because the contractor is expected to meet a rigid set of requirements but is not allowed (or cannot afford) to do custom design. If the government were to select the products and take the risk that the correct match to the user's needs has been made, then industry could implement the design without being accountable, and the risk to industry would be considerably less. Industry should take responsibility for integrating the COTS products well and should not be concerned about whether the product selection was optimum for the user. In my opinion, the government should bear the responsibility for product selection.

A New Development Approach for COTS-Based C^3 Systems

The discussion above results in a different approach to the development of COTS-based C^3 systems, summarized as follows:

1. The Development Command would take responsibility for ensuring that three critical aspects of the development are properly traded off with respect to cost and performance:

   - the balance of the cost to build and modify,
   - the availability of products, and
   - the desires of the user.

2. The Development Command would take responsibility for deciding whether, in a given case, it is better to integrate a system with a single contractor or a multiple contractor arrangement.

3. Even if a single contractor is selected, the Development Command would take responsibility for correct product selection. Contractor selection would be based not on specifications but on the contractor's ability to integrate. Under this new procedure, if the performance of the integrated system is poor, then the accountability would depend on the cause. If both the contractor and the integration plan prove to be sound, then the Development Command would be at fault; if, on the other hand, the integration process was not performed well, then the contractor would be at fault and should be replaced.

4. The Development Command must have a reasonably comprehensive knowledge of the COTS market place to ensure that the final selection of products is picked from the huge number of commercial products that are available. The process requires confidence that a good source has not been overlooked or that some different COTS vendor will not later confront the developer with a superior product.

5. The Development Command must be prepared to evaluate an architecture for basic soundness and for the ability to change at reasonable cost. The user must be convinced that the selected architecture is resilient and adaptable. The government's job is to select the system architecture that satisfies the user's requirements, at a time when the final system requirements are not fully known.
6. Finally, the Development Command must find a way to judge whether the integration activities performed by the contractor were good or bad. Today, this issue is confused by the need to meet specifications. When a high-quality integration effort has been conducted, it is not fair to hold the contractor responsible for the performance of commercial equipment and software.

This summarizes a new model of how COTS-based developments should be done. It is important that an effective approach for this class of development be put in place, because the economics of system development will greatly emphasize the integration of COTS products.

Life-Cycle Issues

We are entering a period in which the DoD will not have enough money to replace C^3I systems. We are going to own systems for a very long time. Much thought has to be given to how we are going to deal with that. Who should maintain these commercially based systems once they are shipped? Industry? The Using Commands? The various Logistics Commands? We have no ready answers, but these are important questions that should not be dismissed on the basis of a given Command's mission or the assertion that the user must retain control of the process.

To modify or make evolutionary improvements to a system composed of COTS products, we should not attempt to modify the off-the-shelf products. The new additions should be more off-the-shelf products, with perhaps some custom-designed elements. The most important factor in upgrading a system should be knowledge of the system's architecture. Given that the system's architecture was properly designed to accommodate change, the people who should be doing the life-cycle maintenance and upgrade ought to be experts in that architecture, so that they can properly exploit the system opportunities for change. It is not clear who these people are. When the developer ships a system to a Using Command for maintenance, the last thing we tell them about is the architecture; when they ship it to a Logistics Command for maintenance, the last thing we tell them about is the architecture. When industry takes it over, the people who do the maintenance are not the designers of the architecture, but rather are those who wrote the final application software.

We are not thinking properly about the life cycle. The issues are quite complex: transfers of command, transfers of responsibilities, transfers of money. This is an area where we must create a more efficient way of doing business.

Additional Issues

The following comments included here are drawn from a letter by Alfred J. Mallette, Major General in the U.S. Army, in which he responded to my early presentation on COTS.

> Many defense systems are properly described as "systems of systems." The Army Tactical Command and Control System (ATCCS) is an example of an integrated system of systems. In such cases, an evaluation of the architecture of any one individual system within the complex would be incomplete without considering its impact on the architecture of the whole.
>
> When we develop measures of merit that serve as the basis of evaluating architectures, it will be important to transition these measures to our education and training institutions.
>
> Many of the architectural features that we consider

desirable will depend upon "standards" that unfortunately will not remain fixed, but instead will evolve as the community gains knowledge and responds to commercial and organizational pressures. Our approach to developing architectures must accommodate this evolution.

We need to distinguish mainstream commercial products that are supported by substantial organizations from "garage shop" products whose industrial base may vanish overnight. At the same time, we must continue to promote the innovation that we expect from small businesses.

The increased use of commercial products in our C^3I systems will inevitably raise concern about the possibility of covert introduction of viruses, Trojan Horses, and other security weaknesses into our defense systems. A rigorous process of analysis and verification must be developed to ensure the integrity of COTS-based systems where the danger is significant but the cost of assured integrity will be high.

The Advanced Tactical Fighter could benefit from advanced tactical communications systems currently under development using COTS software.

Commercial licensing practices may constrain the replication and distribution of software for military systems in the field. It may be necessary to create new practices suitable for DoD.

The Software Engineering Institute (SEI) has data which seem to suggest that the process maturity of the defense industry is greater than that of the commercial industry. If this is the case, then we may be taking a step backward in terms of product quality as we move toward increased use of commercial products. We must deal with this potential problem during the time that industry's maturity catches up with the defense industry.

While DoD's needs do not drive the commercial market, the Defense Department may significantly influence the directions that the market takes. However, the DoD is by no means a monolithic organization, and its potential for influence is usually dissipated through its inability to coordinate and focus. Effective methods to allow the DoD to speak with one voice would be helpful.

Chapter 5
An Examination of Defense Policy on the Use of Modeling and Simulation

Simulation should be a focusing mechanism for running expensive but useful operational tests as early in development as possible.

This chapter is extracted from the first part of the 1990 Defense Science Board report on modeling and simulation; I was a member of the panel and wrote this material in its entirety. Appendix 3 presents the conclusions and recommendations of the panel.

Simulation is becoming a more valuable and widely used tool throughout the Department of Defense. In all three Services, it is now common for development decisions to depend heavily on the results of elaborate simulations. As defense budgets decline, we can expect this trend to continue and perhaps even to accelerate. Political factors also come into play; for example, as the current changes in Europe continue, there is increasingly stronger resistance to the massive war games and battle exercises that used to be held nearly every year. We will undoubtedly be forced to run the games using simulation rather than driving tanks through the German countryside.

As computers become more capable, there will be many new roles for them in simulation. These new roles are not, however, the subject of this chapter — instead, I will limit my discussion to examining the role of simulation in testing and acquisition.

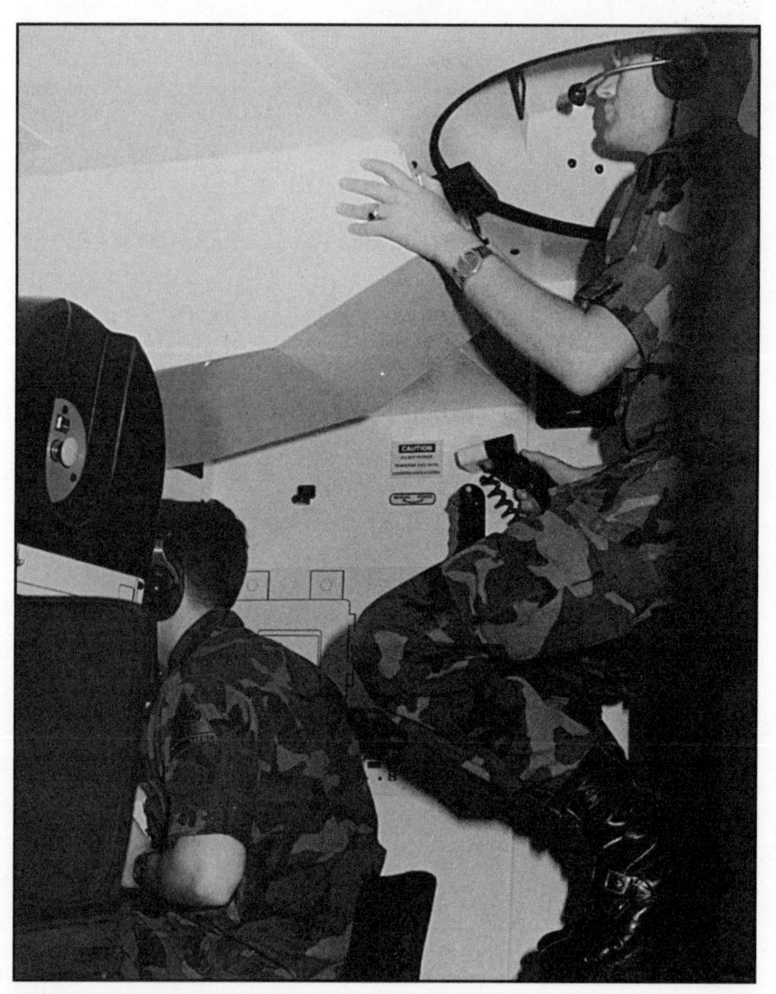

Training for Army tank crews in the SIMNET distributed simulation network. Courtesy of DoD.

Over the past 10 years there has been continually increasing emphasis on the use of special government test organizations (independent of acquisition organizations), whose role is to define, conduct, and evaluate the results of operational tests on newly developed systems. These organizations are used to strengthen the acquisition process by adding confidence to production decisions.

Why has it been thought necessary to create special test organizations? The DoD is concerned that the roles of the different parties in a typical development effort are too rigid and too insulated from the desired overall goal. The user is expected to affirm the utility of the system both before and during its development. The program manager sees that his or her job is simply to develop a system, with no additional requirement to evaluate the user's judgment of utility of the finished system. Thus an independent organization is asked to evaluate the system's utility after it has been developed, as a final check that the process has not gone astray.

A significant side effect of the emphasis on operational testing after development has been a corresponding shift away from its use as a learning tool during development. In former years, the developer commonly conducted operational testing to improve his or her understanding of the complicated interrelationships among system specifications, design, and utility; but testing is rarely done today for this purpose. I believe that this trend is undesirable, and I will include several examples indicating that the consequences are negative.

Operational testing is expensive, and fiscal pressures on the DoD will tend to reduce the amount of testing that is performed. Despite these limitations, it is essential in some cases that operational testing be conducted during development to provide confidence about the utility of the prospective system and to help the developer learn what to do if the tests are unsuccessful. I suggest that this should be the primary role of simulation — to identify and focus on areas of concern related to the ultimate utility of the system under development. When the concerns raised by simulation are sufficiently strong, then operational tests should be performed. Simulation can identify the critical tests and can help provide the justification for the expense of testing.

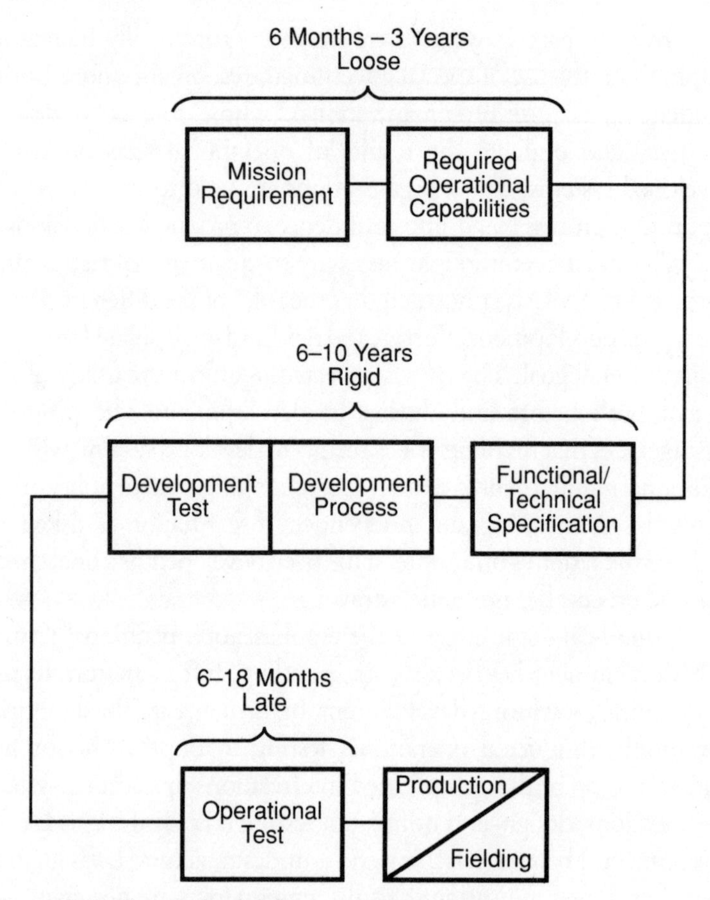

Figure 5-1. The standard defense acquisition process.

However, to be used in this fashion, simulations must be credible to the decision-makers. I will discuss the problem of simulation validation and other aspects of credibility after I say more about the acquisition process and the root causes of operational test failure.

The Acquisition Process

The acquisition process consists of the requirements, development, and operational test phases (refer to figure 5-1).

The requirements phase is the period during which the Services articulate their desire to improve their capability to perform a given mission. In the process of generating requirements, they must combine their experience in military operations with their expectations of future systems technology to bring about some vision of how a new system can add to their mission capability. This involves making assumptions, such as what the threat will be, what our doctrine (and that of the enemy) will be, what the operational procedures for using this new capability will be, what the deployment of the new system will be, and so on.

During the requirements phase, the opinions of the user and development communities can fluctuate significantly, partly as a result of analytical tradeoffs and simulations and partly because the process, by its nature, is imprecise. The debates and shifting opinions

U.S. Army AH-64A Apache Advanced Attack Helicopter firing 2.75 inch rockets during testing. Courtesy of DoD.

are generally not the result of poor or inadequate work — there is simply little reliable information available to make predictions about our own forces, let alone those of the enemy, at this stage.

The development phase begins with the creation of a technical specification for the product to be developed. The specification is generated by two groups: an operational group, which had the original vision of mission utility, and a more technically oriented group that will eventually be responsible for the development activity. By its very nature, it is a process prone to error, but not necessarily human error. The specification-generation process involves even greater levels of prediction and extrapolation, adding to the uncertainty of the loose process upon which it was founded.

Once the specification has been created, it becomes the basis for a contract that must be rigorously managed to specific and binding details. The development process lasts many years: A program manager who does a good job is one who keeps the program stable through rigorous management.

Operational testing is the final phase. At this point an independent group evaluates the system to verify its utility and production worthiness. There are, in fact, three simultaneous evaluations taking place in this phase. One is an evaluation of the equipment itself. The second is an evaluation of the early work done by the planners (who imagined what the utility would be if such a system were built). The third is an evaluation of the process that translated the vision of those operational planners into a specification. While we make three evaluations during operational testing, the last two could have been made years earlier if the system or some approximation of it had been available. In the sense that two of the three evaluations have experienced long and essentially unnecessary delays, they occur very late.

The tests are also late in that the cost of discovering and correcting a problem at this stage of acquisition is greatest. If the program is cancelled, the money expended on development is lost. If a decision is made to correct the system's flaws, the cost is extremely high, because, in addition to the cost of redesign work for the product, there are also all of the support costs associated with changing drawings, changing support equipment, and so on.

When a serious failure is discovered during operational testing, this failure is referred to as a "surprise" — *i.e.*, an unanticipated outcome — because we would surely would not have continued system development for so many years if we expected failure at the end of the process. As everyone knows, such surprises are very unpopular in the Defense Department. In addition to the surprise associated with a particular program, the credibility of the whole acquisition process is called into question. When the process itself loses credibility, it has serious negative effects on everything the Department of Defense does. It is therefore extremely important to avoid surprises, both for the direct cost to the system in question and the credibility loss to the process as a whole.

Unanticipated Results from Operational Testing

Almost all unanticipated results, or surprises, that occur during operational testing can be categorized into four types.

The first type is the change-in-assumption surprise. As I mentioned above, the system planners must make assumptions that lead them to believe the system or technology they want to advance will be useful. These assumptions include the threat, the deployment of the system, the environment it would operate in, and so on. After the typical 10 years of development have gone by, the threat, the deployment plans, the key features, or other basic assumptions have often changed significantly. The operational test discovers that some of these changes are crucial and that the product's utility has decreased dramatically, to the point where production is inadvisable.

A good example is the Division Air Defense (DIVAD) system, a mobile radar-directed anti-aircraft gun. DIVAD was originally conceived by the Army in the early 1970s, and development started in 1977. Its purpose was to protect the moving army from attacks by fixed-wing aircraft (the primary threat) and from stand-off helicopter attacks (the secondary threat). When the program was initiated, the stand-off helicopters were assumed to carry missiles having a range of three kilometers, so the designers of the DIVAD decided on a firing range of four kilometers.

DIVAD development proceeded through 1985. During that time, the threat changed in two ways: The helicopter became the primary threat and the range of the helicopter's stand-off weapon increased to six kilometers. The operational test determined that the DIVAD's firing range was inadequate, given the extended stand-off range of the helicopter threat. The result was that the program was cancelled after a very large investment. The development community was well aware, of course, that the threat was changing, but they argued for many reasons that it was still logical to develop the DIVAD with its four-kilometer firing range. The point is not that we should have produced DIVAD, but that there was no need to wait until the end of the program — and its associated large expenditure — to decide that the change in assumptions was crucial. Perhaps analysis by simulation could have led to an earlier decision.

The second type of surprise is the measures-of-effectiveness surprise. During the requirements and specification phase, the planners must choose some way in which they express utility. Usually they describe a specific task or mission and assert that the system will be considered useful if it can successfully perform the task. During the operational tests following the development process, however, an independent team sometimes runs the test with a different set of measures — different enough so that what seemed a very useful system no longer appears to be useful at all. The result is that the system does not meet the new measures of effectiveness and is cancelled. One particular version of this situation occurs when the original planners say, "If the system under evaluation results in a capability that is better than anything in the field today, then the system will be acceptable to us." However, the unpleasant measures-of-effectiveness surprise arises when (10 years later) a different group of users says, "No, that's not good enough — we demand a specific level of utility, and unless the system can meet this threshold of utility it will not be acceptable at all."

A good example of this case is Aquila, another Army development. The Army currently operates weapons (such as the 155mm howitzer and the Multiple Launch Rocket System) that can fire their ordnance at targets up to 20 kilometers away, yet have no way of detecting targets at that distance. Aquila was a small unmanned aerial vehicle intended to carry sensors that could detect and designate enemy targets many miles away. The vehicle was to have television and infrared sensors and a laser designation system.

In 1974, the original planners stated that the system would have adequate operational utility if Aquila's sensors could detect half the enemy targets in their area of vision and if, when targets were observed, the weapons could exploit the observation and actually destroy the targets 85 percent of the time. In addition, it was required that Aquila should be easy to use in the field. The program continued from 1974 to 1987, when an operational test was run. At that time, Aquila's sensor suite included only the television sensor, and there was confusion during the test resulting from a lack of experience in operating unmanned vehicles. The test was run and the system generally satisfied all the original measures of effectiveness. Nevertheless, the decision makers determined that the system was not good enough to warrant production, and the program was cancelled. Large amounts of money had already been spent.

The test report provided no specific criteria for acceptable utility. How could we have paid for 13 years of development, with the designers and planners having a vision of what was good enough, and then have arrived at a point where it was decided that the system was inadequate? Why were there no substantiated measures that the defense community as a whole had adopted? Perhaps conducting analyses and simulations early in the development phase could have helped. In addition, it is interesting to note that many senior military people still believe we should produce Aquila — in fact, there is nothing currently in development to give extended range to the Army's long-range weapons.

The third type of surprise is the lack-of-maturity surprise. There is always a natural tension at the end of the middle stage of development, when the time has come to begin operational testing. The very first engineering prototypes are available for test, but they are not yet mature. Typically, the software contains bugs and the hardware reliability does not yet meet specifications; these are areas where we must have real-world experience in order to mature. The developer must determine if it is appropriate to take the time to gain the necessary field experience, to allow the system to mature so that it will have a better chance of passing an operational test. The cost of that decision is the expense of leaving a factory idle, a factory that has many workers and machines ready for production. Faced with that cost, the developer may elect to push the product into test prematurely. (The developer's hope is that the product will pass the test, yielding a decision to produce, and getting the factory working as quickly as possible. The maturing process can then be accomplished during the long time it takes to complete the first production units.) Almost invariably, the Defense Department takes the course of a riskier entry into operational tests to gain the economy of rapid production. In most cases this is probably a wise decision; however, on occasion, a test encounters major problems with an immature system.

A good example occurred with the Joint Tactical Information Distribution System (JTIDS), a tri-service airborne datalink system. The objective was to produce a system with a reliability of 400 hours mean time between failures. When JTIDS entered its operational test, the testers noted that the reliability was poor (about 40 hours), so poor that it actually disrupted the tests. They stated, therefore, that the reliability appeared inadequate. This was no surprise to the development community, because they knew that the JTIDS units had not had a chance to mature. But it was a great surprise to members of the three Services who had not been informed about the system's immaturity.

This kind of surprise calls into question the credibility of the organization developing the system and of the management group in the government. The subject then expands from a test of JTIDS to a test of the credibility of the whole acquisition system that created JTIDS. This test does no one any good, and is also very inefficient. Red teams, special panels, and briefings to anyone who has any affiliation with the system are the normal means of recovery, but it is a very long time before credibility is regained. In fact, on JTIDS, during the year or so while all the reviews were in progress, the system matured and eventually showed about 80 percent of the specified reliability in its tests. The program is now back in a more normal mode of development, but at the expense (both in time and in dollars) of a long period of credibility loss, credibility that may never be fully regained in the system.

Developers should share their knowledge of the state of maturity of the system with the full set of involved players, rather than have a large number of people be surprised by a demonstration that the system is not mature. This requires analysis and simulation to aid in determining the projected growth in capability resulting from maturity.

The last surprise is the lack-of-usability surprise. This occurs when, for whatever reasons, the user — the soldier or pilot or sailor who is going to use the system — has no chance to try the system under realistic conditions until the operational test. The surprise results when the users reject the system because it is too difficult to use.

A good example of this is the Strategic Air Command Digital Network (SACDIN). The network was designed to disseminate emergency action messages to our strategic missile force and to receive status messages back from that force. When SAC stood down in June 1992, SACDIN became defunct in name only: the system is still in operation, but is no longer referred to as SACDIN. SACDIN used message-entry terminals comparable to workstations common in offices today, except that these workstations used extensive software measures to ensure security. Because the network carried information critical to our nation's survival, it had to be exceptionally secure; any possibility that some unauthorized party could tamper with the data was intolerable. The system completed a normal development and the operational test was run.

This was the most secure software system of its time; no one had ever gone to such lengths in ensuring security. At times, this led to rather stringent procedures. For example, if an operator entered several incorrect inputs in sequence into a SACDIN terminal, the system viewed this as a potential security breach. The system's response was to freeze the terminal, audit the most recent data, and sound an alarm to bring in a security officer.

During the operational test, the users (who, of course, had no experience with the system) made keyboard entry errors. These errors satisfied the criteria for a potential security breach, freezing the system. At the end of the test, each user said, essentially, "I cannot use this terminal — every time I make a mistake, it freezes instead of helping me." The system failed to pass its test.

The developers made software changes to make the system more user-friendly. However, the process for software change was extremely complicated and it took about a year to revamp the system to solve the problem. As with the original development process, the software's ability to maintain security had to be mathematically verified and there had to be manual validation that the real software matched the specification (for instance, professional teams attempted to break into the system). A complex regression-test was required before the new software changes could be accepted. To avoid such problems, human-machine interaction must be evaluated early in the cycle by use of rapid prototypes and simulation.

Advances in Simulation

The defense community has been using simulation for years, to support virtually every aspect of system development. This includes simulations for operations research (which help developers to understand the utility of a new product in the very early planning stages), simulations that aid in the design of hardware and software, simulations (such as cockpit simulators) that deal with human factors, simulations that synthesize complicated environments (such as enemy jamming and interference) in which systems will operate, to the ultimate war-gaming simulations that are usually the basis for our operational tests.

Over the last 10 years we have seen dynamic growth in the computing and networking technologies that form the underpinning for digital simulation. There has been a corresponding increase in the use of simulation, which can now be both much more elaborate and much lower in cost than in earlier times. This trend will almost certainly continue for the next 10 years.

Since few things get cheaper and better with time, we should seize this opportunity to take more advantage of simulation. In fact, the DoD is giving much greater emphasis to simulation as a tool for

Virtual reality simulations of sub-system components are expected to facilitate training for maintenance and repair of Space Station Freedom.

reducing cost. The following examples illustrate how simulation is changing and what additional value can come from these changes. The examples are drawn from government activities in which MITRE is involved.

One of the best examples is the Warrior Preparation Center (WPC), which is designed to give senior European battle commanders and their staffs the opportunity to train for the operational level of war, using interactive computer simulations that replicate, as closely as possible, the real NATO environment. WPC allows staffs from around the world to participate simultaneously in some of the most sophisticated and realistic war games.

WPC uses parallel-processing algorithms to satisfy the immense processing requirements of these complex simulations. A multiple-user system interface allows for the networking of simulations from many different locations and maintains the integrity of the distributed database. The system offers a faster, more realistic simulation to a larger number of staff members in more widely separated locations than ever before.

For the Strategic Defense Initiative, the Experimental Version Prototype System (EVPS) is being used to simulate and analyze potential strategic-defense systems. It models a set of proposed boost-phase defensive systems against a threat from intercontinental ballistic missiles, where the enemy has the option of using several different launch strategies. The EVPS can simulate all the sensors, weapons, battle management, and communications functions that will be required of any future strategic defense system.

The most interesting aspect of EVPS is the organization of the simulation itself: EVPS is being developed as a structured prototype, where specific goals are established for a fixed number of releases. Each release adds new functions into the model, increasing the complexity and fidelity of the simulation and providing new insights into both the problem being modelled and the model itself.

The National Air Traffic Control Simulation has to simulate one of the most complex problems in existence — the entire air traffic control (ATC) system of the United States. It has to be linked to

Simulation and modeling advances will compress development of systems such as the Tomahawk cruise missile, here on a test flight over the Pacific Ocean. Courtesy of U.S. Navy.

other ATC prototypes and cockpit simulators as well as to training rooms at operational facilities, and development has to continue while the system is in use. A distributed architecture is the only possible answer to these requirements; but that raises several problems, such as database contention (where several users try to read or write the same data at the same time) and reconciliation (making sure that everyone is using the same set of data in the face of constant changes from many sources). These problems are currently the subject of some new approaches.

One approach to achieving high speed is known as time warp. Each of the individual simulators in a distributed-architecture system runs independently; each stores all its previous states in memory. A separate unit "watches" the outputs from each separate simulator and stops them if there is a conflict or if one of the simulators gets too far ahead of the others. Using the stored machine states, it then "rolls back" each simulation to the same time, combines the outputs where necessary, and directs the overall simulation to proceed once more. This technique allows the multiple simulations to race along at maximum speed, stopping and rolling back only when absolutely necessary. The approach is very fast, but requires a large amount of memory and processing capability.

Another approach is the moving time window, in which the multiple simulations are run up to a certain point in time. When the fastest simulation reaches a specified time ahead of the slowest simulation, it is frozen while the slower units catch up. This approach is much less demanding of computer resources, but it is also a great deal slower.

One of the most important recent advances in simulation is called virtual reality, in which the user is totally immersed in the simulation. For example, planners of NASA's space station can enter the details of a space-station module into the simulation — its shape and size, the layout of all the equipment inside, the lighting, and so on. Associated with the system is a special helmet that contains a graphic display faceplate and a sensor to monitor the precise position of the wearer's head. The simulator can then calculate the view that the wearer of the helmet would see if he or she were actually standing in the space-station module, and can change the view in real time as the helmet moves or as the person changes his or her position. This imagery creates a striking sense of reality for the user, allowing a much more intuitive appreciation of the relationships and nuances in the synthesized environment. Recent versions are beginning to allow the user to manipulate objects in the virtual world as well.

These new techniques will likely change simulation completely, but even the most sophisticated simulation will be of no use if confidence in its accuracy is low.

Validating Simulations

Confidence or credibility in a simulation is without doubt the overriding issue. If a system simulation were trusted absolutely by the system developers and all the related decision-makers, no operational tests would be necessary at all. On the other hand, if they had no confidence in the simulation, everything would have to be tested. Obviously these are the extreme cases, but real life often approaches the extremes rather closely.

The following three examples are typical of the difficulties encountered in validating simulations.

Extrapolation

Over-the-horizon radar can serve as an example to illustrate the problem of extrapolation. Over-the-horizon radars transmit high-frequency electromagnetic waves that bounce off the ionosphere and illuminate targets at very long range, well beyond the line of sight. These radars were developed by both the Navy and the Air Force to detect and track large targets, such as bombers that might be attacking the United States.

The performance of high-frequency radar is critically dependent upon the path geometry of electromagnetic-wave propagation, the time of day, the time of year, and solar activity, because the properties of the ionosphere vary appreciably with each of these parameters. To accommodate this variation, the original designers added extra capability to the over-the-horizon radars so that they could maintain performance in the face of abnormal propagation conditions.

While the early systems were being developed, questions arose about the detection of cruise missiles, because the Soviet threat was changing from bombers to stand-off cruise missiles (which have smaller radar cross-sections and are, therefore, more difficult to detect). The Air Force asked whether these radars could reliably detect cruise missiles and, if not, what modifications could be made to the radars to give them this capability.

Six organizations had developed simulation models that they used regularly to analyze high-frequency radar performance against bomber-size targets. Experts in these organizations gave significantly different answers to the Air Force's question. It soon became apparent that the experts were working under different assumptions, so the government set up a special process in which all the experts were asked to estimate radar performance under identical circumstances.

Figure 5-2 shows the results: each of the black bars represents one organization's estimate of the target size (in square meters) for which the radar would achieve a 50-percent detection probability. The answers ranged from 80 square meters to a few square meters.

How is it possible that these experts, using their own simulation models in which they each had confidence, could give such widely different answers? The difficulties arose because, in this application, the models were being extrapolated beyond their regions of validity. Although they gave roughly similar results when the radar target was large, they diverged in their predictions against small targets.

A close examination of the technical details inside the models soon revealed that they all lacked fidelity in accounting for various propagation phenomena, such as ionospheric focusing and multipath, and that they contained subtle differences in their representations of these effects. After much study, a community consensus was reached on appropriate algorithms for each of the important phenomena. The Air Force also calibrated the models by conducting field experiments with an over-the-horizon radar against small airborne drones (facsimiles of real cruise missiles). This work eventually

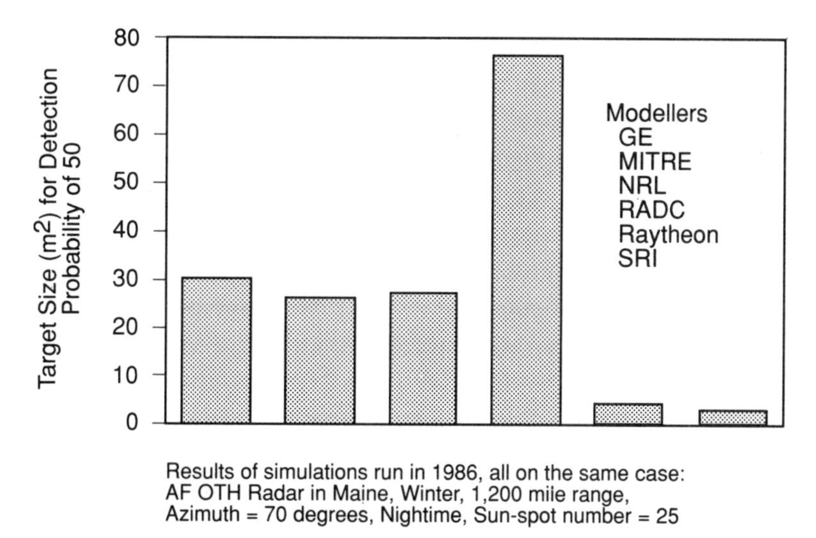

Figure 5-2. OTH radar simulation results.

resulted in a radar model appropriate for small-target use and earned the confidence of the community.

The point here is that confidence in a simulation model requires not only knowledge about the model, but also information about the extrapolation involved in dealing with the specific problem.

Marketplace-Validated Models

The next example is the marketplace-validated model. These are models that companies create and sell and that are used frequently by other companies and the government. The market is the test of validation: If the models work, people buy and use them. Unfortunately, this mechanism is far from perfect.

One illustration is found in response-time models for data-processing systems. When a developer is building a large distributed data-processing system, perhaps a worldwide system with many thousands of users, it is important to know how long it takes for a user requesting data to get a response from the system. One of the key factors is the amount of contention (competition) for computing, storage, and network resources when multiple users request service simultaneously. If there is substantial contention and the management of the processing and communications resources is inefficient, responses from the system can take a very long time.

How does a software designer know what the contention will be in a system that has not yet been put into the field? In this case, there is no way to make measurements, so the designer is forced to make some assumptions about contention. People with experience in this field know that this is very frequently misjudged: as a result, models often produce very wrong predictions. How can this problem be solved? It is extremely helpful to get an early version of the system out into the field, where measurements can be taken, even though the system may be far from complete. Actual data, even when approximate, are generally much preferable to no data at all, and provide at least partial validation of the modeling assumptions.

Reliability Models

Another example is found in reliability models. The problem here is one with which electronics developers have to contend: Given the design of an electronic system, the developer must estimate, before it has been implemented in hardware, how reliable it will be before a development decision can be made. Models can provide some information — they account for the quality of the parts, the thermal stresses, and the electrical stresses that the system will have in operation, and through some integrated set of calculations determine what the mean time to failure will be. These models are used all the time, but most of them account only for electronic parts factors — they do not address factors such as workmanship, manufacturing quality, or even larger parts issues, such as the mechanical rigidity of the boards used or the reliability of the connectors.

Even though these models can provide accurate answers for only a portion of the problem, developers find them valuable because at least they can verify that parts-related factors are not limiting the reliability of the system. However, a developer attempting to validate reliability models must also attempt to measure the quality and workmanship standards for each of the factories.

These three examples illustrate that the concept of validating simulations is extremely difficult.

Simulation in Operational Testing

The Joint Surveillance Target Attack Radar System (Joint STARS) provides an outstanding example of how simulations should be used. Unfortunately, this type of use is not common enough in defense acquisition systems.

Joint STARS is an airborne radar system capable of detecting slow-moving vehicles (for example, tanks on the battlefield), thus serving as a surveillance resource for many Army and Air Force weapons. The radar detects moving targets by directly measuring their velocity and thereby separating them from the relatively stationary ground clutter. (Therefore, the slower the target, the more difficulty a radar has detecting the target against clutter.) A radar of this sort is capable of measuring only the component of a target's velocity projected along the imaginary line joining the radar to the target. This component is called the radial velocity of the target.

Figure 5-3 illustrates the basic technical design issue involved in developing the Joint STARS radar: how slow a target should the radar be able to detect? The figure was derived by creating a model of the roads and off-road areas in Europe passable to tanks and trucks. The modelers laid a hypothetical enemy force of moving vehicles down on this area and calculated the radial velocity of each target to the Joint STARS radar. They then computed the fraction of the target set that would be visible if the radar could detect targets moving at greater than some specified minimum radial velocity. The two curves show the percent of targets seen by the radar as a function of the minimum detectable target velocity, for fast-moving and slow-moving targets. It can be seen that the percentage of slow-moving targets *declines dramatically* as the minimum detectable velocity for the radar is increased.

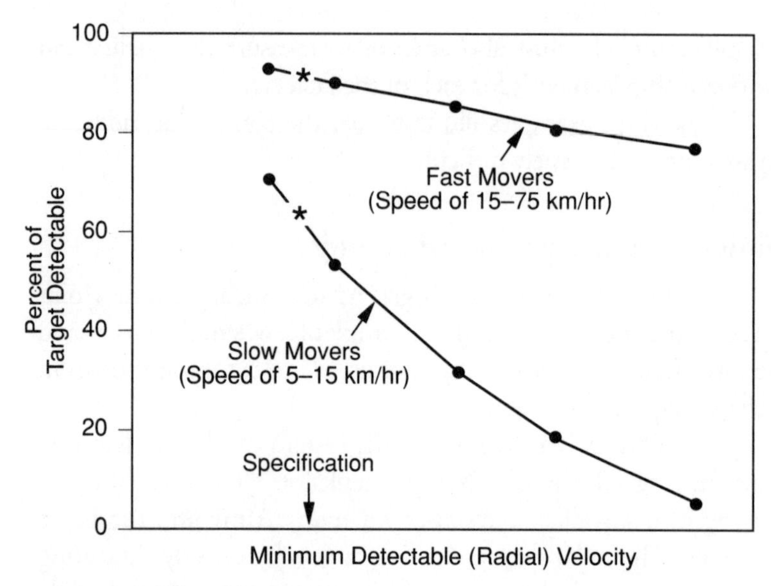

Figure 5-3. Joint STARS operational simulation.

On the basis of these curves, the two Services demanded that the radar be sensitive to very slow targets, and specified the minimum detectable velocity as shown by the arrow (the actual number is classified). If the radar could detect every target that moved at or above this speed, it would detect 95 percent of the fast-moving targets and 70 to 75 percent of the slow-moving targets.

The radar designers performed a corresponding simulation. For a given set of assumptions about the radar's frequency, antenna length, platform speed, clutter behavior, and so on, they estimated the relative ability of the radar to detect a target as a function of the target's radial velocity. They found, as shown in figure 5-4, that the radar's detectability moves down a *very steep curve* below a certain target velocity. The specified minimum target velocity shown in the figure was dictated by the state-of-the-art of radar design.

The designers recognized from these simulations that the estimated performance—and therefore the operational utility—of the system was very sensitive to the fidelity of the simulations. Because the curves shown in figures 5-3 and 5-4 both have very steep portions, small errors in the simulation could result in large changes in system utility. The simulations strongly suggested that more work to increase confidence would be desirable, rather than waiting for 10 years of development to discover whether the system had utility.

Given this focus of attention, an aggressive program of experimentation and early operational testing was established. Component and subsystem tests were conducted to measure antenna performance, oscillator stability, behavior under vibration, and other factors as part of the process of validating many of the internal

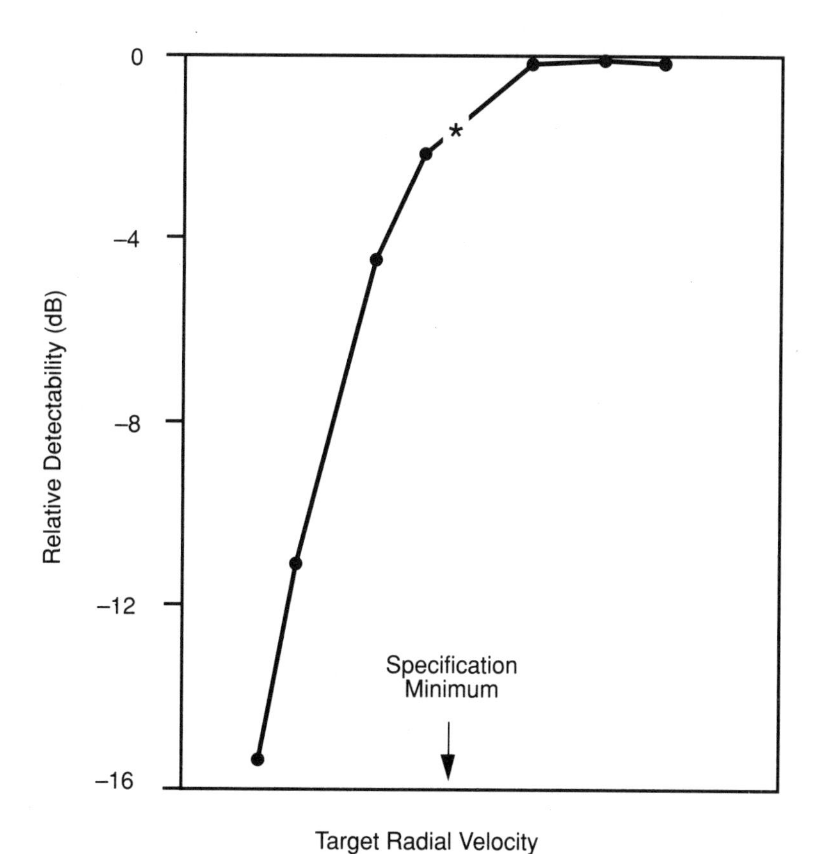

Figure 5-4. Joint STARS radar performance simulation.

assumptions made within the simulations. The system developers scheduled operational flight tests as soon as the basic (but incomplete) system was capable of operation, as another step in getting early validation of the simulation results. An operational test in Europe was conducted to verify additional assumptions about the target, clutter, and interference environment to be expected. At the same time, the designers began to show the data to operational users in order to get from them an early indication of whether they thought the system would have utility at the end-point of development.

Conclusion

It is this concept for simulation that should be stressed—namely, that simulation should be a focusing mechanism for running expensive but very useful operational tests as early in the development process as possible. This is infinitely preferable to waiting six to 10 years for the operational test, which is otherwise the first opportunity for understanding whether the system has utility.

Appendices

Appendix 1: Median and Latest Technology for Selected Systems

This appendix identifies, for each of the systems mentioned in figure 2-15 (Chapter 2, "Modernizing Electronics in DoD Systems"), the dates associated with the earliest, median (the "median" date refers to the design age of the majority of the parts in the system), and latest technology implemented in the design. The acronyms used in the figure are shown below in parentheses in the title of each system description. The components, parts, or integrated-circuit chips associated with the technology eras for each system are also identified.

Automatic Communication Processor (ACP)

The Collins Government Communications Division of Rockwell International developed the SELSCAN processor for the Army and as a commercial product to provide automatic communication-link establishment for high-frequency (HF) radios. The Air Force initiated advanced development contracts in 1982 to militarize the SELSCAN by providing address protection, slow frequency-hopping, etc. The earliest technology is from the related family of HF radios produced in 1968. The median year for technology is 1973, when development was initiated for SELSCAN. The latest technology is from the advanced development work performed in 1982.

AFSATCOM Satellite Terminal (AFSATCOM)

The AFSATCOM ultra-high frequency (UHF) satellite terminal development contract was awarded to the Collins Radio Division of Rockwell International in 1973. The earliest technology is from development activity in 1966. The median technology is based on components identified during the critical design review held in 1974. The latest technology is for the Dual Modem Upgrade, which was initiated by MAC in 1973 to provide the Milstar UHF capability.

AN/ARC-190 HF Radio (AN/ARC-190)

The Collins Government Communications Division of Rockwell International developed airborne HF radios as a commercial product to replace earlier tube-type technology. The commercial radios and a militarized version (AN/ARC-174) were available in 1976. An upgraded radio, the AN/ARC-190, was delivered to the Air Force in 1984. The earliest technology, from 1965, is associated with the power amplifier. The median technology reflects the legacy of components selected in 1970. The year of the latest technology, 1980, applies to the synthesizer.

Ballistic Missile Early Warning System Modernization (BMEWS)

The modernization of all three BMEWS radar sites was initiated in 1974. The upgrade of the site at Thule, Greenland was awarded to Raytheon in 1983; full operating capability was achieved in 1987. The modernization of the site in Fylingsdale, England (1988-1992) used the same technology as at the Thule installation. The site upgrade at Clear, Alaska is currently unfunded. The earliest technology is represented by solid-state modules from 1974. The median technology is from computer technology in 1983. The year of the latest technology, 1984, is associated with the system critical design review for Thule site.

Communication System Segment (CSS)

CSS is the message-distribution system for the national command systems located in the Cheyenne Mountain in Colorado. The initial CSS contract was awarded to Ford Aerospace in the early 1970s. The system has been upgraded over time since its 1973 initial operating capability. The earliest technology is from 1973, for the Data General Nova 840 processors. The median technology is from Ford Aerospace's interface adapters in 1977. The latest technology is from Honeywell DPS 6000 processors in 1985.

Communication System Segment Replacement (CSSR)

CSSR is the replacement system for CSS. The initial contract was awarded to GTE in 1984. The earliest technology is from miscellaneous components from 1978. The median technology is from Stratus XA 2000 processors from 1987. The latest technology is from high-speed circuit monitoring from 1989.

Dual Frequency Minimum Essential Emergency Communications Network Receiver (DFMR)

The DFMR provides separate Ground Wave Emergency Network (GWEN) and Miniature Receive Terminal (MRT) receive-only capability in the same system. It is designed for launch-control center applications and provides substantial space reduction. Westinghouse was awarded the development contract in 1989. The earliest technology in the system is carried over from the MRT (the predominant MRT technology is from 1984). Most of the technology pertains to the digital receiver (1987) that replaces the GWEN and MRT analog receiver functions. The latest technology will reflect component updates using 1990 technology.

Airborne Warning and Control System (AWACS)

The AWACS contract definition was initiated in 1968. The prime contract was awarded to Boeing in 1970, and production began in 1975. The earliest technology is the use of core memory, from 1958. The median technology is the brassboard subset developed in 1973. The latest technology is from color displays from 1984. This information is for currently flying U.S. operational aircraft (the E-3A system) and does not include any planned updates.

E-4 SHF Terminal (E-4 SHF)

The super high frequency (SHF) terminal for the E-4B airborne command post was developed by Rockwell International. RCA developed the antenna and Magnavox redeveloped the USC-28 modem for airborne application. The earliest technology is from the advanced development model program in 1968. The median year for technology is estimated to be 1972. The latest technology is the result of modifications to the USC-28 airborne modem (1980).

Global Positioning System (GPS Receiver)

The GPS satellite system initial operational capability is in 1990, and the final operational capability for the satellites is expected in 1993. The GPS receiver (IA9000 Series) for fighter aircraft was developed by ITT/Plessey. Full-scale development (FSD) commenced in 1978. The earliest technology is from 1978, when FSD started. The median technology date (1983) reflects the components selected to meet the reliability and size goals. The latest technology is from 1990 VHSIC and ASIC updates.

Ground Wave Emergency Network (GWEN)

GWEN was designed to provide survivable low-frequency communications using commercial off-the-shelf equipment. RCA was awarded the Thin Line Communications Capability design-phase and implementation-phase contracts in 1982. The earliest technology in GWEN is from the "motherboards" and the antenna tuning unit that originated in 1967. The median technology is from the Intel 8086 microprocessors used in the RCA-designed equipment (1977). The latest technology was selected before the preliminary design review held in 1982.

Milstar Ground Terminals (Milstar)

Two Milstar Phase I development contracts for satellite ground terminals were awarded in 1983. The Phase II contract was awarded to Raytheon in 1985. The earliest technology reflects the vintage of components selected in 1975 for the ASC-30 terminal advanced development model developed by Raytheon in the late 1970s. Most of the technology is from the critical design review held in 1985. The latest technology was inserted in 1988 for the Modem Processor Unit.

Military Microwave Landing System Avionics (MMLSA)

MMLSA combines the new commercial microwave landing system capability with a modernized version of the old Instrument Landing System (ILS) capability so that the resulting package can fit in the existing ILS space in fighter aircraft. Competitive development contracts were awarded to Rockwell-Collins, Plessey, and Hazeltine in 1989. The earliest technology in the system is from the 1985 processor. The median technology is from the application-specific integrated circuits from 1988. The latest technology is from the Monolithic Microwave Integrated Circuits (MMICs) inserted in the 1990s.

Miniature Receive Terminal (MRT)

The MRT was designed to provide receive-only communications for the B-52 and B-1 bombers via the Minimum Essential

Emergency Communication Network. The MRT program upgraded and miniaturized the capability deployed in launch-control centers in the 1960s and 1970s. The MRT development contract was awarded to Rockwell in 1984, with a production contract awarded in 1990. The earliest technology is from standard parts (1980) selected by the contractor (analog receiver front end). The median technology date, 1984, is represented by the signal-processing portions of the system. The latest technology reflects refinements in signal-processing technology from 1986.

North Warning System (North Warning)

UNISYS was awarded the contract to develop the AN/FPS-124 minimally attended short-range radar for the North Warning System in 1984. The production contract was awarded in 1990. The earliest technology is from solid-state radar components in 1975. The median-technology decisions (*e.g.*, large-scale integrated circuits) were made in 1984, and the latest technology is from the application of micro-strip technology in 1990.

Over-the-Horizon Backscatter Radar (OTH-B Radar)

The Air Force initiated advanced development of OTH-B radar in 1970. The OTH-B experimental radar contract was awarded to General electric in 1975, and the contract to upgrade the experimental site to the first operational sector was awarded to General electric in 1982. The earliest OTH-B radar system technology is represented by tunable high-powered high frequency (HF) transmitters from 1970. The median technology is from DEC VAX distributed-computer technology in 1982, which is the initiation of the full-scale development contract. The latest technology is from 1990, intended for incorporation into the Alaskan OTH Radar System.

PAVE PAWS 1 and 2 (PAVE PAWS)

The PAVE PAWS system consists of four phased-array radars placed within the continental U.S. for the detection of submarine-launched ballistic missiles. The development contract for PAVE PAWS 1 and 2 was awarded to Raytheon in 1976, and the contract for sites 3 and 4 was awarded in 1983. The earliest technology is from the Rome Air Defense Center effort to develop phased-array technology in 1965. The median year for technology is 1975; this was the year that the decision on a solid-state receiver front end was made. The latest technology is for the processors from 1977, marking the date of the critical design review. These technology dates are for the radar. The technology of the automated data-processing equipment and the modernization of sites 3 and 4 is somewhat later (1983 to 1986). Sites 1 and 2 have been upgraded (1988 to 1991) to provide system-wide data-processing commonality.

Rapid Execution and Combat Targeting (REACT)

REACT provides centralized monitoring and control of higher-authority digital communications systems, message processing, communications integration, and delivery of error-corrected Emergency Action Messages for ballistic missile launch-control centers. The full-scale development contract was awarded to GTE in 1989. The earliest technology is for visual display units and keyboards from 1985. The median technology is from 1988 (*e.g.*, the MILVAX computer). The latest technology is for radiation-hardened components from 1990.

SAC Digital Network (SACDIN)

SACDIN was a modernized subsystem of the Strategic Air Command's (SAC's) Automated Command and Control System. SACDIN became defunct in name when SAC stood down June 1, 1992. The systems are still in operation, no longer under the name of SACDIN. The development contract started as SATIN IV in 1976. The contract was restructured between 1977-1978 to reduce scope and became SACDIN, with ITT as the prime contractor. The earliest technology in SACDIN is the IBM Series I processor from 1970. The median technology is represented by decisions made at the system design review in 1977. The latest technology is for floppy-disk drives for the mass storage system (1979).

Space Defense Operations Center (SPADOC 4)

SPADOC 4 is being implemented at Cheyenne Mountain and the off-site test facility in three phases by Ford Aerospace. SPADOC 4A implemented the earliest technology (Megatek work stations and IBM 3083 mainframes, which dated from 1983) and the median technology (Micro VAXs from 1988). SPADOC 4B provided additional IBM mainframes (3081 and 3090) that are from 1988, the median year. The latest technology from 1991 is implemented in SPADOC 4C, which installed DEC VAX 6000 Series computers for the communications front end.

UHF Satellite Terminal System (USTS)

USTS is being developed by MAC for Military Airlift Command (now replaced by Air Mobility Command) aircraft. The full-scale development contract was awarded in 1987. The earliest technology is for processor chips from 1982 that reflect the legacy of MIL SPEC standard parts. The median technology is semi-custom large-scale integrated circuits for the demodulator and dual-ported memory chips from 1987, which is the year the preliminary design review was held. The latest technology is for custom very-large scale integrated (VLSI) circuits from 1988.

Appendix 2: The Strategic C^3 Budget

This appendix estimates the value of the strategic C^3 "plant" over the last 15 years (FY78 to FY92). This is the strategic portion of the total budget for C^3 investment. C^3 investment consists of the following C^3 expenditure categories: research, development, test and evaluation (RDT&E); procurement; and military construction expenditures.

The budget data provide a component breakdown (RDT&E, procurement, military construction, operations and maintenance (O&M) and military personnel) and a functional breakdown (strategic, theater & tactical, and defense-wide).

Since the data for FY78 and FY79 were not conveniently available in the desired form, we extrapolated the FY80 data backwards to obtain estimates for those two years. (The proportion of C^3 investment in FY80 to the total DoD budget in FY80 was used.)

The total estimate for C^3 investment is $124 billion in then-year dollars, as shown in table A-1. [The data for FY80 through FY89 represent budget authority; the data for FY90 and FY91 are budget requests.] To estimate the strategic portion of C^3 investment, we computed the strategic C^3 expenditure to total C^3 expenditure (31%) and applied this to the C^3 investment total. The estimate for total strategic investment is $38 billion (in then-year dollars) or $49 billion (FY92). This was rounded to $50 billion (FY92) in the discussion in chapter 2.

Table A-1. Total C^3 Budget Investment Over Fifteen-Year Period

Cumulative Total	FY78 to FY92 (Then-year dollars)
RDT&E	$45B
Procurement	$77B
Military Construction	$2B
C^3 Investment	$124B

Appendix 3
Defense Science Board Panel Recommendations*

The point was made that simulation is generally not something that confirms or rejects a hypothesis, but is instead a mind extender — it makes us think about an area of concern that we would not have focused our attention on otherwise. As a result, simulation can lead into areas of evaluation that may be crucial but that might not have been considered.

It was also noted that central management of the reuse and distribution of simulation ignores a very important point: Simulation designers themselves carry all the knowledge about what went into the models and about what can be extrapolated and what cannot. Transferring the models without the designers would be an error, because many of these programs are so large that it would be impossible for another organization to understand all the subtle ingredients.

The Learning Role

Imagine setting up, at the start of a development program, measures of effectiveness, assumptions (such as threats and environments), and an understanding of how testing and simulation will augment each other over the life of the program. There is no doubt that most of these measures and assumptions will be inaccurate at the outset, but imagine further that they can be continually refined as knowledge is gained throughout the development planning process. We will call this an evaluation framework.

Our first recommendation is that we need to emphasize the learning role for operational testing during development; the panel

---

*This appendix presents the recommendations and conclusions of the 1990 Defense Science Board Panel on Modeling and Simulation. In the Defense Science Board report, these recommendations follow the material presented in Chapter 5.

would like to see this standardized by the setting up and documentation of evaluation frameworks at the beginning of every program. This will also provide assurance that the program will not encounter the change-in-assumptions surprise (as DIVAD did) or the change-in-measures-of-effectiveness surprise (as Aquila did). To do this, we have to involve the independent operational test people from the start — they cannot be brought in at the end of the process.

This raises two concerns. The first is that the testers will not remain independent if they get involved in the programs earlier than they do now. The committee's view is that aloofness should not be confused with independence. The value of independence is that the testers have knowledge to provide and a management chain that gives them the ability to apply it; to keep them aloof is to lose this.

The second is a concern about how these operational testers can properly lay out the evaluation framework, since they are a small group and do not have the resources or knowledge to do it. We believe that the development community (not the testers themselves) should take the lead in laying out an evaluation framework, and that the community of operational people should be involved in agreeing to and continuing to modify the framework as development goes on.

Simulation to Provide Focus

As noted earlier, simulation cannot, in most cases, prove or disprove hypotheses, but it can isolate high-sensitivity areas, areas that could change the prevailing view of system utility. We want to establish an important role for simulation in performing excursion analyses to focus the early operational tests. The second recommendation of the panel is that the Department of Defense should require sensitivity analyses at the beginning of all development programs. We do not want to see fixed-point simulation results; we want excursion analyses that can be used as the basis for deciding whether early operational testing should be done. In other words, simulation should focus, not replace, testing.

Periodic Re-Evaluation

The third recommendation deals with the rigid period of the development cycle. Simulation is an evaluation tool, and if we have a management process that attempts to keep contracts and specifications fixed, then there will be no room for evaluation (because the purpose of evaluation is to determine whether the specifications are correct).

The panel believes that the current acquisition process stifles evaluation. Program managers are not motivated to examine their programs—they are interested simply in stability. The culture in the Defense Department should promote evaluation. The panel recommends that the Office of the Secretary of Defense (OSD) establish policy and provide guidance to the acquisition community for systematically re-evaluating system specifications using modeling, simulation, and testing.

This is a very difficult recommendation, and we do not have a simple solution. Nonetheless, we think this is crucial if we are going to get any value from the tools and capabilities of simulation. This raises another concern. Some will say that this will encourage a mode where we are always changing everything, and that we will end up with nothing (in other words, we will lose management control). In response, we want to make clear that we do not advocate giving up configuration management on system developments, nor do we insist that every evaluation should result in a change. There should be two distinct processes: one that is evaluating and one that is changing. We expect the rate of evaluation to be much higher than the rate of change. However, if we do not evaluate at all, there will be no changing, and then we will end up facing one or more of those surprises.

Human Factors

The fourth recommendation is stimulated by the human-factors problem illustrated by the now-defunct SACDIN. Because the tools are now available and the cost is sufficiently low, every program should build mock-ups of human-machine interfaces as soon as possible, and bring in the real users to get a better grasp of the design's utility. The panel recommends that the service acquisition executives ensure the use of human-in-the-loop simulations in all development programs, beginning with requirements definition and continuing throughout the acquisition process.

There is one set of systems where this is particularly difficult: the large command and control systems used by many geographically separated generals or admirals. It is not easy to bring them to a central facility to test the system. The solution is a new technique called distributed simulation, where many decision makers can play their parts in a global simulation while remaining in or near their own offices. Perhaps through that mechanism, we can get higher-level Defense Department staff members to test those portions of the systems that they will ultimately use. We are seeing the technique used today for training, but we think that it can be useful in the area of development as well.

Credibility

Finally, the panel made several recommendations with regard to the issue of credibility. How do we know whether to trust a simulation? Should we set up a central office to accredit simulations? Should we set up a management process to distribute and reuse simulations?

The panel felt that there are no single-point answers to the problem of trusting simulation. We should be doing excursion analyses and sensitivity analyses; when we find something that makes us nervous, we should run a test — the test, and not the simulation, validates the answer.

We should have professionally documented simulation results. Decision makers should see the whole set of data so that they understand how a simulation was calibrated and to what degree the results are dependent on extrapolation.

Should we set up a central office? The panel believes that we clearly should not — there is no office with the capability to perform this difficult task. Should we set up a distribution process for distributing these simulations? The panel believes that we should not — if we cannot accredit simulations, we certainly cannot have a very logical process for distributing them. But the panel does have some recommendations concerning validity.

There are certain models in the Defense Department that tend to be used and re-used by expert groups — for example, the Defense Nuclear Agency (DNA) models on nuclear effects or the Defense Intelligence Agency threat models. The DNA models have never been fully validated, and they often lead to problems. At the same time, they are the best available, and everyone uses them. Models of this type should have a budget line to reinforce their improvement and keep them current. We recommend that the Joint Chiefs of Staff and OSD allocate money directly to those groups that develop and use the models.

When an acquisition decision depends heavily on simulation, an independent panel may be used to perform several valuable validation tasks. While it is nearly impossible to validate an entire simulation, a panel can validate many aspects of it, including the people who designed the simulation. It can also validate the extrapolation (that is, it can ask specific questions about the historical use of the simulation versus the extrapolation now being used) and the fidelity of the input data. The panel can determine if users are doing a partial evaluation or a full evaluation, and, in the case of a partial evaluation, whether the parts not being evaluated are important to credibility. These are all questions that panels of experts can answer in a fairly short time, to add at least that level of confidence to the use of simulation when needed. Such panels, however, should be used only in special cases.

In regard to professional documentation, Defense Acquisition Board documentation has a place for, but does not specifically call for, the data that determine the basis for validating the simulation and the credibility factors. We believe that these data should become part of the documentation.

Finally, as the over-the-horizon radar example points out, strange results sometimes arise when comparing different validated models against the same problem. Such comparisons, when available, can be valuable indicators of the overall credibility of the models.

Summary

The panel believes that it is extremely important to avoid operational test surprises, for reasons of both cost and credibility. This can be done by performing more operational testing during development, enabling developers to learn about problem areas while they still can be fixed rather than waiting until the system is put into the field. It also helps to develop evaluation frameworks at the onset to specify how evaluation is to be performed; as the real program progresses, these must be upgraded so that they are consistent with the state of knowledge. Operational testers must be involved so that there are no unanticipated gaps or changes in viewpoint.

The current acquisition process stifles evaluation, and unless we have a more open attitude to performing evaluation, the problem will remain unsolved. Something must be done at upper management levels to change the process, but this does not include an independent simulation office to accredit or manage the use or distribution of simulations. Such an office cannot add confidence, but it can add confusion. Finally, the panel believes that simulation should be used to focus testing onto those areas where we do not have confidence.

Glossary

Author's Note:
The U.S. Air Force activated new commands called Air Mobility Command (AMC, *q.v.*) and Air Combat Command (ACC, *q.v.*) on June 1, 1992. Military Airlift Command (MAC), Strategic Air Command (SAC) and Tactical Air Command (TAC) were deactivated at the same time.

ABCCC: The Airborne Battlefield Command, Control, and Communications System is an airborne platform that carries commanders, operators, and a large variety of radios to link with multiple Service elements during theater actions.

ACC: Stands for Air Combat Command, which activated June 1, 1992, when both Strategic Air Command (SAC) and Tactical Air Command (TAC) deactivated.

ACP: The Automatic Communications Processor is a component of the Air Mobility Command's high-frequency (HF) radio system that automatically selects the best frequencies from an assigned set and automatically establishes communications links.

Ada: Ada is a high-order computer language that is said to offer greater productivity, fewer coding errors, and better prospects for reuse of program modules. DoD now directs that Ada be used for all military software.

ADM: ADM is an acronym that means Advanced Development Model.

AFSATCOM: The Air Force Satellite Communications System, predecessor of Milstar (*q.v.*), consists of satellites and ground terminals that operate in the UHF frequency band.

AFSATCOM Modem: The modem is part of a ground terminal that provides UHF satellite communications to the AFSATCOM space segment and the UHF portion of the Milstar system.

AMC: Stands for Air Mobility Command, which was activated June 1, 1992 when Military Airlift Command was deactivated.

AN/ARC-190: The ARC-190 is a high-frequency (HF) airborne radio system featuring selective calling, preset channel scanning, link-quality analysis, and automatic frequency selection.

AN/SRC-16(A): The SRC-16(A) is a shipboard high-frequency (HF) radio communications system.

AN/TRC-170: This is a family of digital troposcatter radio terminals developed for use by tactical ground forces.

Aquila: Aquila is a small, unmanned, airborne vehicle that carries sensors for detection and designation of enemy targets many miles from the ground-control terminal.

ATCCS: ATCCS is an acronym that means Army Tactical Command and Control System.

AWACS: AWACS is the Airborne Warning and Control System. It is a sophisticated airborne surveillance radar system, using a rotating antenna housed in a mushroom-shaped radome atop the aircraft, with complete facilities for controlling air operations.

B-2: The B-2 is the "Stealth" bomber, designed for low radar cross-section and long range.

Berlin ATC System: The Berlin Air Traffic Control System is an automated traffic-control system incorporating improved computers and consolidated radar displays. It is located at Templehof Airport in Berlin.

BMEWS: The Ballistic Missile Early Warning System is a network of three strategic radars, located in Alaska (Clear), Greenland (Thule), and England (Fylingsdale). They are designed to provide early warning of ballistic missile attack against continental United States. All three radars are tied into the North American Aerospace Defense Command.

C^3I: C^3I is an acronym that means Command, Control, Communications, and Intelligence. It includes surveillance and intelligence sensors, command centers that may contain decision aids for the commanders, and communications to link the sensors with command centers and intelligence analysis centers.

CCPDS-R: This term means Command Center Processing and Display System Replacement. It is part of the complex command-and-control system located in Cheyenne Mountain, Colorado.

C.I.S.: This is an abbreviation for Commonwealth of Independent States, formerly the Soviet Union.

Cobra Dane: Cobra Dane is a phased-array intelligence radar system that observes and collects data on missile-system test flights. The radar is located in Shemya, Alaska.

Cobra Judy: This is a shipboard surveillance and tracking phased-array radar used to collect intelligence data on ballistic missile tests.

COTS: COTS is an acronym that means Commercial Off-The-Shelf. It is used in referring to hardware and software that are commercially available. The reference to "off the shelf" implies that the commercial items have been thoroughly tested and are mature, *i.e.*, that the risks in incorporating them into a system are low. See also GOTS.

CSS, CSSR: CSS is the Communications System Segment, which is the communications network for the air defense, space defense, and missile warning systems at the North American Aerospace Defense Command's Cheyenne Mountain Complex (located in Colorado). CSSR is the upgraded replacement system for CSS.

DART: DART stands for Detection, Action, and Readiness Tracking System; it is used by the Naval Sea System Command to identify systems not meeting operational goals.

DCA: This is the former Defense Communications Agency, which is now called Defense Information Systems Agency (DISA).

DFMR: DFMR is the Dual Frequency MEECN Receiver; it is a low-frequency/very-low-frequency (LF/VLF) receiver that receives the LF ground-wave emergency network (GWEN, *q.v.*) signals and the VLF Minimum Essential Emergency Communication Network (MEECN, *q.v.*) signals. DFMR, MEECN, and GWEN are all part of a system that ensures communication during nuclear attack.

DISA: DISA is the Defense Information Systems Agency (formerly Defense Communications Agency, or DCA).

DIVAD: The Division Air Defense system was designed for the Army to be a mobile radar-directed antiaircraft gun.

DODIIS: This acronym means Department of Defense Intelligence Information System.

E-4 SHF Terminal: This is a communications terminal that operates at Super High Frequency (SHF); it is used in the E-4 Advanced Airborne Command Post.

EMP: EMP stands for Electromagnetic Pulse, which is a phenomenon associated with nuclear detonations. The high-voltage pulse can destroy or disable many kinds of electronic equipment over distances of hundreds of miles or greater.

Emulation: Emulation is an approach that allows software designed for old computer hardware to run on new hardware, whereby the new hardware "emulates" the old hardware as exactly as possible.

EVPS: This acronym stands for the Experimental Version Prototype System, used to simulate and analyze potential strategic-defense systems that are part of the Strategic Defense Initiative.

FAA: FAA means Federal Aviation Administration, the government organization responsible for aviation safety and air traffic operations in the United States.

Firmware: This term refers to computer-program instructions that are stored in read-only memory (hard-wired or field-programmable ROM) rather than in general-purpose read-and-write memory. The instructions are "firm" because computer programmers cannot easily change them.

GIITS: GIITS is the General Imagery Intelligence Training System, a computer-aided system that provides instruction in imagery interpretation, analysis, and reporting for intelligence analysts.

GOTS: GOTS stands for Government Off-The-Shelf; it is used in referring to hardware and software developed specifically for the U.S. government, when the development is considered complete and mature enough to be used with little risk in some other program(s). *See also* COTS.

GPS: GPS is the Global Positioning System, a satellite-based navigation system that provides both military and civilian users with extremely accurate position information.

Granite Sentry: This is a command post, located in the Cheyenne Mountain Complex in Colorado, for the assessment of ballistic-missile warnings from our strategic radar and infrared sensors.

GWEN: GWEN stands for Ground Wave Emergency Network, an extremely low frequency (ELF) communications network designed to provide survivable links between national command authorities and elements of U.S. strategic forces in the event of nuclear war.

Have Quick: Have Quick is a jam-resistant ultra-high frequency (UHF) voice radio system designed for U.S. Air Force tactical operations.

Have Sync: Have Sync is a U.S. Air Force very high frequency (VHF) jam-resistant airborne radio, capable of both voice and data communications and compatible with the U.S. Army ground radio system.

IOC: IOC stands for Initial Operational Capability or Initial Operating Capability.

ISO: This acronym stands for the International Standards Organization.

Joint STARS: The Joint Surveillance Target Attack Radar System, also known as Joint STARS, is an airborne phased-array radar system (including command and control operators) that is capable of detecting slowly moving ground vehicles such as tanks and trucks.

JTIDS: The Joint Tactical Information Distribution System is a sophisticated digital radio communications network that provides for jam-resistant secure exchange of data among Army, Air Force, and Navy units.

LINK 11 and TADIL-A: These are digital data formats used to exchange target track information and other data.

LVFE: This acronym stands for Low Volume Force Element, a small, light-weight communications terminal for use with the Air Force Milstar satellite communications system.

MAC: MAC was the abbreviated name of the Military Airlift Command, which is now replaced by Air Mobility Command (AMC, *q.v.*).

MEECN: This term stands for the Minimum Essential Emergency Communications Network, intended to provide secure communications just before and during a nuclear exchange. *See also* DFMR, GWEN.

Message Handling System: The Message Handling System receives messages on a network and automatically disseminates them on the basis of pre-arranged formats and message "profiles."

Microcode: Microcode is a form of low-level software that resides in a special portion of certain computers, a configuration that can lead to greater efficiency and speed of execution.

Milstar: Milstar is a satellite constellation that uses extremely high frequency (EHF) radio to provide the military Services and the National Command Authority with secure, instantaneous worldwide communications, even during a nuclear conflict.

Mission Planning System: The Mission Planning System is an automated aid to assist in the planning of air missions. It helps the operator and pilot to optimize the intended flight path with respect to enemy threats, attack profiles, visibility, weather, etc.

MLS: The Microwave Landing System is a lightweight mobile instrument-landing system that can be deployed quickly on bare landing sites or on bases where equipment has failed or has been damaged.

MMIC: MMIC (sometimes referred to as MIMIC) is an acronym that means Monolithic Microwave Integrated Circuit. MMIC chips are miniature microwave circuits that are generally fabricated using Gallium Arsenide materials.

MMLSA: This acronym stands for Military Microwave Landing System Avionics. It refers to the miniaturized electronic packages carried aboard military aircraft that operate in conjunction with the microwave landing system to be employed at many military airports.

MRT: MRT stands for Miniature Receive Terminal; it is a very-low-frequency (VLF) receiver used on airborne platforms to receive MEECN signals. *See also* MEECN and DFMR.

NIST: NIST is the National Institute of Standards and Technology, formerly the National Bureau of Standards.

NORAD CMC: NORAD is the North American Air Defense Command. Its headquarters are in the Cheyenne Mountain Complex (located in Colorado), a sophisticated worldwide system responsible for the aerospace defense of the North American continent.

North Warning System: The North Warning System is a strategic warning system consisting of 14 long-range and 39 short-range advanced radar systems located in northern Alaska and southern Canada. The radars are designed to detect and track aircraft and cruise missiles.

NORTIC: NORTIC is a command post designed to help NORAD track drug smugglers.

OSD: OSD stands for Office of the Secretary of Defense.

OTH-B Radar: Over-the-Horizon Backscatter Radar is a sophisticated long-range radar system developed by the Air Force to detect and track airborne targets approaching North America. It operates in the high-frequency (HF) band and "bounces" its signals off the ionosphere to achieve detection beyond the line of sight.

Patriot: Patriot is a defensive missile system developed by the Army to intercept aircraft. It was hurriedly modified in the Gulf War to intercept Scud missiles from Iraq.

PAVE PAWS: PAVE PAWS is a set of four long-range, phased-array radar systems designed to provide warning of attacks from submarine-launched ballistic missiles against the United States.

Porting: Porting is a process by which old software is re-hosted or moved to a new computing platform.

R&M: This abbreviation means Reliability and Maintainability.

REACT: The Rapid Execution and Combat Targeting system is an upgrade to the communications and launch-control capabilities of the Strategic Air Command's Minuteman intercontinental ballistic missiles.

Red Team: This is an informal term for a team of experts who are brought in to review and assess a program that is experiencing some sort of difficulty (schedule or cost overruns, performance shortfalls, etc.), and to suggest "fixes" or remedial actions.

RFP: This acronym stands for Request For Proposal.

SACDIN: The Strategic Air Command Digital Network was an improved data-communications system that provides high-speed, high-quality information interchange to intercontinental ballistic missile bases in the United States. It has, since June 1, 1992, become defunct in name. The systems are still operational, but are no longer called SACDIN.

Scud: A ground-to-ground missile system used by the Iraqis in the Gulf War.

SEI: SEI is the Software Engineering Institute, part of the Carnegie Mellon Institute, located in Pittsburgh, Pennsylvania.

Sentinel Bright: This system provides the Air Force Air Training Command with computer-based training for cryptologic linguists and electronic system intelligence operators.

Sentinel Byte: Sentinel Byte is a system whose mission is to make intelligence data, originating at higher command levels, available to mission planners.

SHF: SHF stands for Super High Frequency.

SINCGARS: The Single-Channel Ground and Airborne Radio System is an FM-band radio system that provides survivable jam-resistant voice and data communications between U.S. Army ground troops and U.S. Air Force aircraft.

SPADOC: The Space Defense Operations Center is the integrated command, control, communications, and intelligence fusion center for space-defense missions. It is located in the Cheyenne Mountain Complex in Colorado.

Tri-Tac: The Tri-Service Tactical Communications System is a secure digital data-communications network using fiber optic and radio connections designed to link Army, Air Force, and Navy forces.

Unix: Unix is one of several standard operating systems for digital computers.

USTS: The Ultra-High Frequency Satellite Terminal System will provide the Air Mobility Command with interoperable communications for moving military personnel and equipment throughout the world.

VHDL: VHDL stands for VHSIC Hardware Design Language. It is a computer-aided tool that facilitates design and analysis of systems, ranging in size from a microelectronic chip to an assembly such as a board or a cabinet, and — at least in principle — even to a full-scale C^3I system.

VHSIC: VHSIC stands for Very High Speed Integrated Circuits, referring to microelectronics chips designed under the government-sponsored VHSIC program and meeting high performance and density specifications.

WPC: The Warrior Preparation Center uses interactive computer simulations to replicate large-scale war situations. In recent years, it has been modified to allow interactive participation by commanders located around the world.

X-Windows: This is a standard format and protocol that allows distributed computers to interactively display graphics and data in the form of "windows."

Bibliography

1. Skurka, J. M., Livecchi, S. G., "Technology and Efficiency: Partners in C^3I Logistics," *Army*, pp. 56-58, April 1990.

2. Arthur, L. J. *Software Evolution, The Software Maintenance Challenge.* New York: John Wiley & Sons, 1988.

3. Card, D. N. *Measuring Software Design Quality.* Englewood Cliffs, N.J.: Prentice-Hall, 1990.

4. Day, R. *A History of Software Maintenance for a Complex U.S. Army Battlefield Automated System, Proceedings of the Conference on Software Maintenance.* New Jersey: IEEE, 1985.

5. DOD-STD-2167A, Defense System Software Development, 1988.

6. Lehman, M. M., and Belady, L. A. *Program Evolution — Processes of Software Change.* New York: Academic Press, 1985.

7. Rock-Evans, R., and Hales, K. *Reverse Engineering: Markets, Methods, and Tools.* England: Ovum, Ltd., 1990.

8. United States Air Force Scientific Advisory Board. *Report of the Ad Hoc Committee on Post-Deployment Software Support.* U. S. Government Printing Office, 1990.

9. National Institute of Standards and Technology (NIST), *Application Portability Profile (APP), The U.S. Government's Open

System Environment Profile OSE/1, NIST Special Publication 500-187.

Sources Notes:

Chapter 2

A. Carroll, R.T., Russell, R.G., Wickwire, K., Wigler, D.J. *A Procedural Example of Miliary Electronics Systems Modernization*; MITRE Technical Report (MTR 11304), January 1992.

Appendix 2

B. *Defense News*, 30 October 1989.

C. Defense C^3I Budget Symposium, AFCEA, 2 March 1989.

Figure 2-5

D. Paraskevopoulos, D.E. and Fey, C.F. *Studies in LSI Technology Economics III: Design Schedules for Application-Specific Integrated Circuits*; IEEE Journal of Solid State Circuits. Vol. SC-22, No. 2. (April 1987): pp. 223-229.

Figure 3-2

E. Arthur, L.J. *Software Evolution, The Software Maintenance Challenge*. New York: John Wiley & Sons, 1988.

Index

D

E

J

L

M

R

S

V

W

X